AGRICULTURE
ABRIDGED

AGRICULTURE ABRIDGED

RUDOLF STEINER'S 1924 COURSE

Jeff Poppen

edited by Hugh Lovel

First edition 2021
Published by Jeff Poppen
Red Boiling Springs, TN 37150
barefootfarmer.com

Quotations and drawings by Rudolf Steiner

Books by Jeff Poppen

The Best of the Barefoot Farmer
The Best of the Barefoot Farmer II
Lessons from Old Agricultural Textbooks

Books by Hugh Lovel

A Biodynamic Farm
Quantum Agriculture

Cover photos of the author's farm and the authors
by Alan Messer

Dedicated to
Hugh Courtney and Hugh Lovel

Contents

Dedication	v
Foreword	1
Introduction to Biodynamics	3
Excerpts from Steiner's Report	9
Lecture One	12
Lecture Two	18
Lecture Three	24
Address	29
Lecture Four	31
Discussion	36
Lecture Five	40
Discussion	48
Lecture Six	51
Discussion	57
Lecture Seven	59
Lecture Eight	65

Discussion   71
Lessons From Old Agricultural Textbooks   77
Afterword   121

Foreword

Soon after I began an organic farm in the 1970's, I got my dad's copy of Rudolf Steiner's Agriculture Course. I immediately realized two things. The sentences and ideas I understood made a lot of sense to me and seemed extremely important and relevant. What I read and did not understand made no sense to me at all. So, I read it, again and again. I was told I didn't have to believe in it for it to work, so I began practicing what is now called the biodynamic method.

In 1987 Hugh Lovel and I formed the southeast biodynamic association with Harvey Lisle and Hugh Courtney. We read the lectures at every annual conference. I became familiar with Steiner's unique terminology and a few more sentences were clarified. His later lectures to scientists, doctors and teachers kept shedding light on the Agriculture Course, which was given shortly after the introduction of synthetic nitrogen as an alternative to traditional farming practices.

When you are enthusiastic about something you want to share it, but it wasn't easy trying to explain this book. I kept notebooks full of my favorite sentences and wrote them in my own words. In 1993, the Creeger-Gardner translation came out, and it really helped. My friendship with Hugh Lovel blossomed into many fruitful farming discussions, resulting in his offer to help edit this booklet. He loved this project, felt that it was needed, and was still working on it when he passed away in August 2020. His introduction is an excerpt from his book, Quantum Agriculture.

Steiner's lectures were given in German and translated from shorthand reports unrevised by him. The George Adams translation was used for this project, and the quotes are from it. I have taken great liberties, just using what I felt I could convey to someone unfamiliar with Steiner's work. My aim is to inspire you to read the Agriculture Course yourself, and put this wisdom into practice. The drawings are by Rudolf Steiner.

In order to understand how farmers kept their farms fertile before 1900, l have included an appendix, Lessons from Old Agricultural Textbooks, from my study of farming books written around the turn of the previous century.

The Agriculture Course, regarded as the beginning of the organic food and farming movement, lays the foundation for a truly healthy agriculture. Steiner's guidelines are the cheapest and easiest way to farm that I've encountered. I'd like to see them reach a wider audience. I hope your interest in the biodynamic method gets aroused, and that you will dig deeper.

Jeff Poppen, December, 2020

Introduction to Biodynamics

by Hugh Lovel

Biodynamic agriculture—the oldest of organic methods—is a science based, holistic, regenerative agriculture that works with life processes to achieve self-sufficient, quality production of delicious high-vitality goods. It grew out of the insights of Rudolf Steiner, whose agriculture lectures at the estate of Count and Countess von Keyserkingk near Koberwitz, Poland in 1924, addressed the shortcomings of chemical agriculture in a truly comprehensive way.

In his twenties Steiner's scientific training was in maths, chemistry, and biology at the technical institute of Vienna. He later earned his Doctorate in Philosophy with his treatise, The Philosophy of Freedom, which advanced the proposition—now accepted as proven in Quantum Physics—that observer and phenomenon are inseparably linked. Clearly the choice of what we look for depends on our concepts, without which we have no grasp of what our senses encounter.

Hired by a publishing house to edit the scientific writings of German literary giant, J.W. von Goethe, Steiner was inspired by

Goethe's explanations of the processes behind physical, measurable occurrences. Each measurement is fixed at a time and place, but over time living things keep changing. Without continuous measurements of these changes no processes emerge. Goethe noted that the butterfly in a museum was only a corpse, and the processes that animated it were missing. Yet, elusive though they might be, these processes contribute enormously to the sense of reality, particularly with living things.

Steiner, who was clairvoyant, investigated folklore, herbal medicine, Eastern religions, native cultures, homeopathic medicine and many other disciplines to acquire the concepts and vocabulary to make sense out of his impressions of nature. Particularly in the last years of his life, his medical and agricultural lectures conveyed an all-encompassing approach to understanding the maths, physics and chemistry of how living organisms function and how their problems can be dealt with. Most of his agriculture course focused on life processes. Ever practical, his remedies for agriculture took into account the environment in the broadest possible sense—the rhythmic motions of the sun, moon, and planets relative to the earth in the context of the universe. His agriculture lectures convey a profound grasp of how processes within living organisms relate to our surroundings.

One of the keys is viewing each property as a self-contained organism, something alive, whether a large farm or a small garden. Biodynamics also clarifies our relationships with cosmological cycles and the activities of nature at large.

Starting with the horizontal lime and vertical silica processes, on the grand scale the life activities of an agricultural operation function between these two axes. This fundamental concept correlates with night and day, winter and summer, nitrogen and carbon, legumes and grasses, soil and atmosphere, inner and outer planets, sedimentary and igneous rocks, plant reproduction and food production. The night-time processes associated with lime and nitrogen relate to mineral release, nitrogen fixation, digestion and nourishment, all of which occur within the soil or work downward into the soil. On the other hand, the daytime processes associated with silica and carbon relate to photosynthesis, blossoming, fruiting, and ripening. These processes arise out of the soil and work through plants into the air.

Understanding biodynamics helps us see the relationships of various agricultural processes to each other, where things fit into bigger pictures and what the causes and consequences of various interactions may be. When we know these things we have better idea of what our resources are and how to guide agricultural events. Then we can build soil as we save time and money, reduce inputs and use what is in nature. In short, biodynamics is a way of making sense of what happens in nature so we eliminate waste and bring things into a healthy, dynamic balance.

Ever the biochemist, in his Agriculture Course Steiner indicated the roles of various elements. He described sulphur as what the spirit 'moistens' its fingers with to work into the physical, oxygen as the carrier of life, hydrogen as the vehicle for the spirit, nitrogen as the carrier of consciousness and carbon as the basis for physical form. By providing a vocabulary for these concepts and showing their relationships to the activities and substances of the world around us, Steiner outlined a framework for creating a much wiser and more rewarding agriculture.

Biodynamics shows us how to make the land thrive, and how to remedy conditions we wish to change. Most of what we need is free—a gift of water, carbon dioxide and nitrogen from the atmosphere. Inputs like lime, gypsum, phosphates and trace elements are remedies for depleted land that has fallen out of balance. Were it thriving it wouldn't need inputs. However, most land today is sick, and we must supply whatever is missing before it can thrive.

Given the necessary ingredients, biodynamics has a toolbox of compelling preparations to impart the life processes needed to draw what nature freely provides into living activity.

For example, silicon is abundant in the soil while nitrogen is abundant in the air. Both are like God, ever present and immediately at hand—but chemical agriculture fails to draw silicon and nitrogen into biological activity. Instead it relies on inappropriate inputs that stimulate without nourishing—fertilizers that reduce fertility. Instead, biodynamics uses small amounts of specially prepared catalytic preparations to engage silicon and nitrogen, which are key for quality production.

Steiner also pointed out that hydrogen, the first and most universal element, was the vehicle for spirit, the prime mover. Humans embody this spirit as ego, a formative force of individuality and self-awareness. Humans, who are aware of their individuality, take a hand in their future evolution; while animals simply go along with the flow that affects them from the surrounding planets and stars. Each species receives its formative forces from a different starry direction.

Biodynamic agriculture crosses scientific disciplines and expands the frontiers of science into realms where life processes and consciousness are generated. At one time this would have been dismissed as imponderable, mystical or, perhaps, delusional. Nonetheless the use of biodynamic methods—especially the biodynamic preparations—keeps increasing since biodynamics gets quality results.

Biodynamic practice is an individual affair subject to complex variables, and results are not guaranteed. Despite many guidelines, the responsibility for success lies with the enthusiasm and clarity of intent of each person applying the method.

We might keep in mind that from the viewpoint of quantum physics, the observer and the phenomenon are inseparably linked. The act of observation is a key factor in determining the phenomenon. Stated in a more time-honored way, what we seek, we find, or what we think, we grow.

Excerpts from Steiner's Report

June 20, 1924 [1]

After returning home from delivering the Agriculture Course, Steiner gave a report about what happened. Lectures began at 11:00 am and lasted until 1:00 pm, and were followed by a mid-day meal and roaming around the grounds of the estate. Many thanks were given to all the workers and hosts who organized the event.

"With regard to the agriculture course, the first consideration was to outline what conditions are necessary in order for the various branches of agriculture to thrive. Agriculture includes some very interesting aspects—plant life, animal husbandry, forestry, gardening, and so on, but perhaps the most interesting of all are the secrets of manuring, which are very real and important mysteries...during the last few decades the agricultural products on which our life depends have degenerated extremely rapidly."

[1] *Creeger-Gardener translation*

Manuring in Steiner's time meant how we fertilize the soil for growing crops. He was well aware of what was causing the degeneration in farm products. It was soluble artificial fertilizers. The secrets of manuring he refers to are the living interactions within the soil. He could discern the detrimental efforts of NPK (nitrogen, phosphorus, potassium) fertilizers on soil bacteria and fungi, whose activity in compost, manures, and soils he referred to as very real and important mysteries. They were mysteries because so few people back then understood microbial soil interactions.

The importance of diverse balanced microbial activity is now well established. We need our microbial partners in our stomachs and on our skin. Animals need them, plants need them, and the soil needs them, too. Their presence in the soil helps ensure healthy plants grow there. An immediate halt to chemical fertilizing and the use of compost would turn degeneration into regeneration. Steiner understands the "significance of the manure for the fields, and why it is indispensable in certain areas, and how it should be handled. No one realizes today that all the mineral fertilizers (NPK) are just what are contributing most to the degeneration of the products of agriculture."

Where nitrogen comes from matters. The nitrogen in the soil must be formed under the influence of the entire heavens. This nitrogen must be alive. There are two kinds of nitrogen, the dead nitrogen in the air above ground and the nitrogen below ground, which is alive. Atmospheric nitrogen is an inert gas, the "dead nitrogen." It is triple-bonded to itself and quite difficult to break apart. Lightning can do it, and so can life processes. Bacteria in the soil, feeding on plant root exudates, can tap into atmospheric nitrogen, and the protozoa that eat them excrete nitrogen as amino acids. This living nitrogen feeds healthy plant growth. "Under the influence of the entire heavens" refers to the whole universe, from the stars, sun, and planets down to the animals, plants and microbes.

Plants get their nitrogen from the digestive activity going on around them, but animals internalize this digestive process. Most animals are microscopic, but just because we don't see them doesn't mean they are unimportant. These microbes are in animal digestive tracts where dead nitrogen changes to living nitrogen. When Steiner says "people fertilize scientifically now," he is referring to the recent introduction of synthetic nitrogen salts. This kind of nitrogen kills off many of the most important soil microbes and promotes the wrong kinds. "No nitrogen fertilizers, no nitrogen salts" and "the only really healthy fertilizer is cattle manure" make Steiner's view of fertilizers crystal clear.

Lecture One

June 7, 1924

After gratefully appreciating the hosts of the Agriculture Course, the first thing Steiner points out is that those who talk about agriculture should have a sound basis in it, and really know what it means to grow beets, potatoes or corn. He includes the social aspects, the organizational nature, and the economic realities.

The social aspects of agriculture are mentioned first, and they echo Tolstoy's[2] observation that the people doing the work are of the utmost importance. The social life in rural areas develops out of families. Everyone knows each other and their peculiar talents, habits, and personalities. This allows for an equitable distribution of work and goods because it is all on such a small, intimate scale. Agriculture and civilization grew up together and remain inseparable. In several passages, Steiner describes himself as a peasant, and honors the wit, observational skills and instincts of country people. He would far rather listen to a chance conversation with a farmer than a scientist.

History suggests that when peasants move to cities, they bring

---

[2] *Leo Tolstoy, 1828-1910, Russian novelist and social reformer*

with them an immense practical intelligence. But removed from agriculture, education tends to lose its sound basis. How rural society takes care of itself is best left to those who live and work there. Humbleness, compassion and practical sense become ingrained in anyone who cares for land, plants, and animals. Social aspects are generally kinder in the country, and best left to those that live there.

The same is true of the organizational nature of a farm. It seems obvious that those who are in constant touch with the land are the ones who best know what to do. It lowers quality, happiness, and health for non-farmers to organize farms. Organizing for short term profit rather than long term abundance creates problems on farms. Living organisms are organized, and a comprehensive view is required to understand the many interactions that make a farm thrive. The farmer is the only one who is in the position to organize this complexity.

The economic realities in farming also need to remain in the hands of the growers. They are the ones who know how much it costs to grow the animals or crops again. Too much interference by others is always paid for by the farmers. Supply and demand can create price fluctuations that don't reflect the costs of production, which, first and foremost, must be covered. Once the farmer is fairly compensated, then and only then should others concern themselves with the price of farm products. Steiner insists, "One must work in a businesslike, profit making way, or it won't come off."

Steiner's affinity with Goethe[3] surfaces when he mentions influences coming from the entire universe affecting what people erro-

---

[3] Johann Wolfgang von Goethe, 1749-1832, German poet and dramatist

neously consider to be self-contained entities. A prerequisite to understanding the agriculture course is the realization that all things in nature are interconnected. A plant cannot be understood in isolation, removed from its interrelationships in the field. It would be like trying to understand why a compass needle points north without taking into consideration the earth's magnetic field.

The interworking of nature is the study of farmers. Centuries of observation have led to crop and animal rotations, the utilization of the various species, and the secrets of building fertility. Farmers' instincts were quite specific and reliable when they were closely connected with nature.

Next, Steiner indicates what is most important in agriculture. Instead of talking about the chemical and physical components of something, he asks us to look carefully at how human beings live. We find a considerable degree of independence from the outer world, but this is less so with animals. Plants are still embedded in, and quite dependent on, what is occurring in their earthly surroundings, which is a reflection of what is occurring in the universe. Plants directly reflect the seasons, in animals it is less so, and humans appear quite free in regard to these cycles.

"In the old instinctive science wherein the sun was reckoned among the planets, they had this sequence: Moon, Mercury, Venus, Sun, Mars, Jupiter, Saturn." These were the seven lights in the sky that moved differently than all the other stars. Planets meant wanderers. Steiner will speak of this planetary life which is connected with the earthly world.

We need to consider the extremely important role silica plays. Silica, a combination of silicon and oxygen, makes up half of the earth's crust. Quartz, sand and many rocks are primarily silica, and so are computer chips. Why is silica so important? Silica helps the communication of information in plants and soils.

Fungal hyphae, the underground parts of fungi, are silica rich tubes that transport nutrients and fluids. They unite the plant root with the soil at large. When a plant needs nitrogen, or any element, a signal is sent down to the root. Bacteria and fungi living on the roots help the plant get what it needs in return for the plant's products of photosynthesis, which feed them. It is a symbiotic relationship in a healthy soil. Every plant species has specific microbes that colonize their roots and ensure healthy growth, and this is facilitated by silica.

Not all nitrogen is the same. Live nitrogen in the form of amino acids is the form most easily used by plants. This comes from living beings in the soil. When the plant needs nitrogen, soil life can supply it. Dead nitrogen, such as nitrates, ammonia or urea, interferes with the functions of silica and beneficial nitrogen fixing microbes. It takes more energy for a plant to use dead nitrogen than amino acid nitrogen. This uses up plant sugars at the expense of both energy and flavor. It's the plants with high sugars that taste better and resist insects and diseases. "No nitrogen fertilizer."

Sunshine striking the leaves creates tension that draws calcium, minerals and amino acids from the soil, through silica rich fibers up into the new growth. Photosynthesis brings carbon dioxide into the leaf during the day, forming carbohydrates with water and giving off oxygen. At night, as the tension relaxes, the carbohydrates sink down into the soil, feeding the microbes that help the plant grow. These microbes require lime, a combination of calcium and oxygen, for nitrogen fixation and growth.

Lime and silica are the great polarities in nature, with the plant in between. The clay/humus complex in the soil mediates the forces of these two poles. Silica makes the plant receptive to the wider context of nature. It even conducts the influences of Mars, Jupiter and

Saturn, the outer planets. Conversely, lime makes the plant receptive to the reproductive influences of the Moon, Venus and Mercury, the inner planets.

The sun travels through space with a vortex of planet orbits following behind it. Steiner postulates that forces from the universe enter the vortex through the outer planets' orbits, and are absorbed by the siliceous rocks in the earth, and then ray upwards. Other forces come from the universe by way of the inner planets and are received by the earth as lime forces raying downward.

Warmth promotes the forces of silica. Plants need warm weather to ripen their nourishing fruits and seeds. A lack of silica will mean less nutrition. Silica is everywhere in minute doses, and fungi have even been found in outer space. Keep in mind that plants directly reflect the stars, and silica is the means for communication of information. When plants become nourishing food or fodder, substances like silica are involved. Both the silica coated, fungal hyphae in the soil and the silica rich tubes of the fibrovascular bundles in the plant enhance the plant's ability to efficiently access a wide array of nutrients throughout the soil.

On the other hand, water promotes the forces of lime. It is the ideal substance for the distribution of lunar forces. Lime brings nitrogen into the plant. Plant growth shoots up after a rain and a full moon. A lack of water or calcium limits the capacity for growth and reproduction, as does a lack of the other related cations, which are potassium, magnesium and sodium.

Steiner notes that people go through life quite thoughtlessly today, glad not to have to think about such things. They conceive of nature in a materialistic way, functioning like a machine. Steiner acknowledges our achievements with machines, and that instinctual peasant wisdom had to step aside for the rise of scientific discovery. But now it's time to join the two world views together. Life does not work in a mechanical way. A living organism is not a simple reductionist system, but a very interdependent interaction of many different things. It reaches out from stars to microbes. This is the primary lesson in the first lecture.

We have come to a starting point with the revelation of how silica and lime work to bring nitrogen into our crops in the proper way for maximum animal and human nutrition. Materialistic thinking is directly responsible for the simple fact that Steiner could not find potatoes as good as the ones he ate as a boy. He tried them everywhere. "Many things have declined in their inherent food values, notably during the last decades."

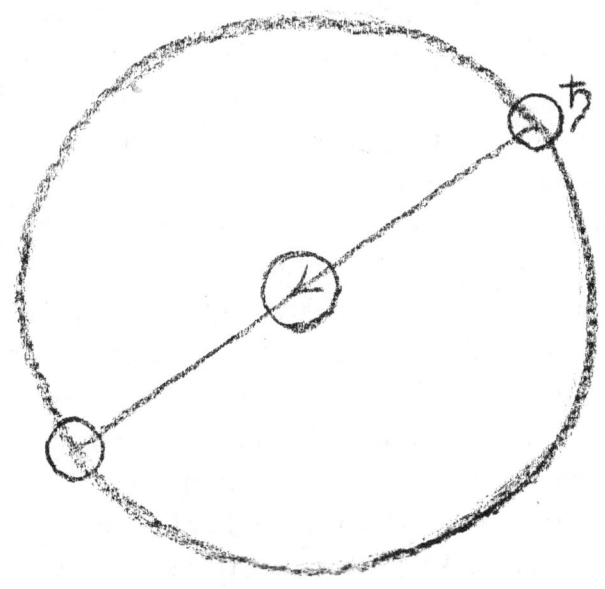

Saturn-forces work both when shining down upon the Earth and when passing upwards through the Earth

Lecture Two

June 10, 1924

Steiner begins the second lecture by giving an overview of the whole agriculture course. We will spend the first lectures gathering knowledge so as to recognize the conditions on which the prosperity of agriculture depends, observing how all agricultural products arise and how agriculture lives in the totality of the universe. Then we will draw the practical conclusions which can only be realized in their immediate applications and are only significant when put into practice. But for now, we must gather, recognize and observe.

Here we see Steiner's holistic and observational approach to science, compared with Newton's[4] reductionism. We are not starting with a problem and hypothesis, as in Newtonian Science. Instead, we are looking for information, conditions, and how something (agriculture) lives. There are no boundaries to where we will look.

The first condition is clear. A farm is true to its essential nature if it is conceived as a kind of individual entity in itself—a self-contained living entity. Whatever you need for agricultural production, you should try to possess it within the farm itself, including in the

---

[4] *Sir Issac Newton, 1642-1727, English mathematician and philosopher*

farm the due amount of livestock. Steiner insists that a farm needs livestock, and then explains why. Our livestock should get their feed from our farm, and our farm should get its fertilizer from them. It is not the same whether we get manure from the neighbor or from our own farm. What makes our manure different?

The humus content of soil, with its carbon, bacteria, fungi and protozoa, is formed in large part by the animals on it. They eat what grows on the farm, and then digest, transform and integrate the farm's microbiology inside their stomachs. A cycle of rejuvenation happens as it is returned to the soil and new plants grow there. He justifies this need for a farm's own livestock by considering, not only the earth but also influences from the distant universe. This digestion by the livestock balances the digestive lime forces of the manure in the earth with the silica forces of information from afar. This will be considered from various standpoints.

We will begin with the soil. The soil is alive and is more than the sum of its parts. We can observe that this invisible life of the earth and its fine and intimate activities is different in summer than in winter.

We can also observe that the surface of the earth is a kind of organ like the human diaphragm. All the plants, animals and people live in the belly of the farm organism. But the head and nervous system of the farm are underground in the roots and soil life, while its reproductive organs are up above, like an upside-down person. There is a continual, dynamic interaction between what is above the earth's surface and what is below.

We are asked to observe what these influences are and where we find them. In lecture one, we were introduced to lime and silica, and their relationship to the inner and outer planets, respectively. The life processes and reproductive activities associated with lime occur above the earth and depend on the inner planets and the sun. The informative and nutritive activities of silica that occur beneath the earth's surface depend on the outer planets and the sun. To better picture how this works, if the soil was just lime the plants would be horizontal and rounded like a cactus. If the soil was all silica, plants would be vertical and thin like a vine.

We are still just gathering various items of knowledge. Notice that both sets of planets work with the sun. It is through the silica of sand, rocks, and stone that we have influences from the distant universe. This is where life comes into the soil, through the communication of information by silica.

You may wonder how what is absorbed by the soil gets carried back up into the plant. The greater surface area and charge of clay particles facilitates this absorption and transportation. Adding clay to a sandy soil and adding sand to a clayey soil are old-time farming remedies because the soil needs both. Clay is the carrier of the upward stream of silica's activity beneath the soil.

Plant growth through photosynthesis in the air above the soil is a kind of outward digestion, not unlike the inward digestion of animals. A mutual interaction arises when what happens above ground is drawn downward by the limestone in the soil. Farmers spread lime on top of their fields, knowing it will work its way downward.

Steiner tells us there are two kinds of warmth for plants, a leaf and flower warmth that is dead, and a root warmth which is living. The moment warmth is drawn into the earth, also by the limestone, it is enlivened. Air, too, is alive in the soil and dead above. The soil is full of aerobic, live beings, much more so than the air.

Earth and water, on the other hand, become more dead in the soil than above it. By losing life they become receptive to distant forces, especially in mid-winter when the minerals in the earth come under these influences. These are the crystal forming forces.

Before and after this period, minerals ray out forces that are particularly important for plant growth. Experiments have shown that crystals precipitate out of solution more easily underground in winter than in summer.

To till the soil, we must know the conditions which enable distant forces access to the earth. We can learn this from the seed forming process. When the plant matures and its protein is the most complex, the lime and silica forces separate and the plant is driven into chaos. In the formation of the seed it becomes receptive to the influences of its surroundings.

We prepare the soil for the seed by creating the condition for a new complexity, by growing cover crops, applying compost and minerals, and gentle tillage. Steiner says that a plant is always the image of some constellation, which is carried forward in the seed. This new organism is formed from the life forces of the entire universe, and the best way to grow the new plant is to place it in a humus rich soil.

In most plants we have the silica nature in the root and the lime nature in the canopy. Normally, the distant silica forces work upward from within the earth in a single, central root up the stalk to the flower. But in a highly divided root, as with grasses, the earthly nature is working downward from its normal place above the soil. These plants form a thatch with their extensive root system and are the fodder plants which build good soil. The best soils are found in the steppes, plains and savannas where grasses have been grazed periodically for centuries. We mimic this by rotationally grazing grassy pastures and sowing grain cover crops. These are all silica rich plants with sharp, pointy leaves.

If we want the silica forces to remain below, we put the plant in sandy soil. Clay helps transport silica forces upward. So, if we want potatoes to form in the soil, and not have the plant shoot up into seed formation, we would rather have a sandy soil.

Steiner encourages us to heighten our observational powers. The lime forces work in the horizontal, as seen in the sedimentary layers of limestone, and the silica forces work in the vertical, as seen in the sharp, pointed mountain ranges. "We can trace the process quite exactly. We can see this directly," he says. Lime works with growth and reproduction, leaves and flowers, and silica works with the ripening of seeds and nutrition.

We must always be able to distinguish between silica and lime forces. Steiner then explains that in ancient times humanity created the different kinds of crops from primitive varieties by this kind of knowledge and instinctive wisdom. We must rediscover this and gain new knowledge of how these things work. For example, people of today do not know that silica receives light into the earth and makes it effective, while the lime and humus in the soil work in darkness. We need such knowledge again.

Different places on the earth have their own specific types of animals. If on any farm you have the right amount of animals to eat the farm's plants, these will give the farm the right amount of manure. Just as Steiner describes the farm organism as a person standing on their head, we must also include how the animals fit in.

The animals live in the belly of the farm organism and we can see the influences of the planets in the forms of animals. From the animal's nose towards its heart, the distant planet forces are at work. In the heart itself the sun is at work, and from the tail towards the heart the inner planets show their influence. You have the true contrast of the sun and distant planets in the form and figure of the animal's head, and the moon and inner planets in the form and figure of the animal's rear. Knowing this you will be able to discover a definite relationship between the manure of different animals and the needs of the earth where those animals graze. The animals which eat the farm's plants will in turn provide manure based on this fodder. Consequently, it will provide the very manure which is most suited for the soil which grew those plants. The farm is healthy in as much as it provides its own manure from its own stock.

Studying skeletons of mammals in a museum can help us learn to read the forms of animals. As we learn to read nature's language of forms we will perceive all that is needed by the self-contained farm organism.

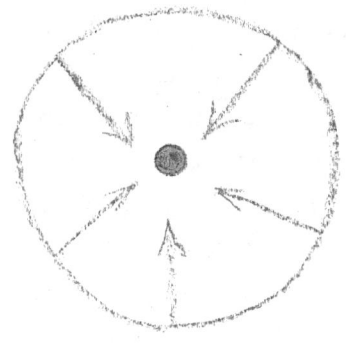

*In the chaos of the seed the new organism is built
up again out of the whole universe*

Lecture Three

June 11, 1924

The third lecture starts off with the question, how do the earthly and distant forces work through the substances of the earth? The significance of nitrogen in all farm production is generally recognized, but the essence of its activity has fallen into confusion. We must look into the wider activities of nitrogen in the universe as a whole. Nitrogen, in living nature, has four sisters that combine with her to form protein. These are carbon, oxygen, hydrogen and sulfur.

Sulfur, a trace element compared with the others, acts as the mediator between the formative power of the spiritual[5] and the physical. Sulfur and phosphorus got their ancient names as "light-bearers" because they carry light into matter. What chemists know of these substances in their laboratories neglects their inner significance in the workings of nature, much like what you learn of some-

[5] *Spiritual, an intelligence conceived of apart from any physical organization or material*

body from their photograph compared to meeting and getting to know them.

Carbon is the bearer of all living forms in nature. It is in all things that are alive or were alive, including coal and diamonds. Through photosynthesis, lifeless carbon dioxide from the air becomes part of the living plant. Animals exhale carbon as carbon dioxide, which allows them mobility. Carbon provides the framework for the other sisters to move through the world. Carbon has the capacity to create the most varied and sublime forms. Life on earth is based on carbon.

Every living thing is permeated by oxygen. Oxygen is alive inside of plants, animals and humans, and when it is in the soil. Science can only see dead oxygen, the fate of considering it only from the physical standpoint. This is the dead oxygen in the air. Oxygen depends on nitrogen to find its way along the paths mapped out by the carbon framework.

Nitrogen has an immense power of attraction for the carbon framework, but carbon can draw nitrogen away from its other functions in the soil. An example is mulching with fresh sawdust, which causes a nitrogen deficiency in nearby plants. Inert nitrogen is a corpse in the air, accounting for 78% of the atmosphere. It only comes alive in living beings, many of which live microscopically in the soil. Nitrogen becomes alive in the earth, as does oxygen. But nitrogen is extremely sensitive, and this is of the greatest importance for agriculture. Nitrogen provides the basis for conscious awareness (intelligence) of the distant forces that work themselves out in the life of plants, animals, and the soil. Nitrogen provides the intelligence that guides the oxygen (life) into the carbon (form). "You can penetrate into the intimate life of nature if you can see the nitrogen everywhere, moving about like flowing, fluctuating feelings." We shall find the treatment of nitrogen infinitely important for the life of plants. In agriculture, above all, no nitrogen salts should be used.

Hydrogen is the first element in the periodic table, and the smallest. It is also by far the most plentiful element in the universe. There must be a constant interchange of substance between the earth and the universe. In living structures, the spiritual becomes physical with the combination of these five elements. All that is living in the physical forms upon the earth must be led back again into the great universe. It is not the spirit that vanishes, but what the spirit, carried by sulfur, has built into the carbon framework, which has drawn life to itself from oxygen with the aid of nitrogen. This must be able to vanish into the universe again. With hydrogen the physical flows outward and is carried away, once more by sulfur. Hydrogen carries the spirit and, using sulfur, connects with carbon, which provides the physical form. Oxygen carries life and nitrogen carries sentience. When the organism dies, hydrogen, again using sulfur, carries out into the far spaces of the universe all that was formed, alive and sentient. So, we have these five substances, each inwardly related to a specific function.

Steiner then asks us a question, "What are you really doing when you meditate?" In meditating you always retain in yourself a little more carbon dioxide than normal and do not drive all of it back to the nitrogen in the air. If you knock your skull on a table, you'll only be conscious of your own pain. But if you rub against it gently you'll be conscious of the table's surface. So it is with meditation. By and by you grow into a conscious living experience of the nitrogen all around you. All becomes knowledge and perception. By meditating, farmers make themselves receptive to the revelations of nitrogen and will farm accordingly. All kinds of secrets that prevail in a farm and farmyard emerge. An uneducated farmer walking over the fields is a meditator and acquires a method of spiritual perception. "The merely intellectual life is not sufficient, it can never lead into these depths."

These five substances, provided by the atmosphere and united together in protein, are bound to other substances and are not independent. There are only two ways they become independent. Hydrogen carries them out to the universal chaos or hydrogen drives these fundamental substances into the protein of seed formation. In seeds there is chaos, and in the far circumferences there is chaos.

For anything to grow, these five sisters need other substances, and these come from the earth. When carbon comes from the air into a plant and then into an animal or human, it must build on a content and framework not only of lime, but also of silica. Limestone and silica come from the earth, and are the minerals that anchor plant growth. Just as our lungs crave oxygen, the limestone in the earth craves nitrogen. The atmosphere, with its nitrogen and oxygen, must find a way into the soil, and one way it does this is through leguminous plants.

The whole organism of the plant world is separated in two when seen from the viewpoint of nitrogen. When we encounter legumes, we are seeing a kind of nitrogen breathing as if by the lungs, and with any other plants we are looking at the remaining organs, which breathe in a far more hidden way and have other specific functions. Legumes are essential in the farm's crop rotations.

Limestone, which drives this in-breathing of nitrogen, has a hungry nature, wanting to draw everything to itself, especially nitrogen. All that limestone craves lives in the plant nature. This must be balanced by an uncraving principle that desires nothing for itself, which is the insoluble silica nature. Like our sense organs which perceive their surroundings but not themselves, silica perceives the surrounding universe. Limestone provides the universal craving, and clay mediates between the two.

Lime underground disturbs carbon, so carbon, the form creator in plants, allies itself with silica to overcome the limestone nature.

The plant lives in the midst of this process, with the limestone clutching below and the silica leading growth upwards. But in the midst, giving rise to all the plant forms, carbon orders all these things. Humus is so important because that's where the carbon is.

Now the question arises, "What is the right way to bring the nitrogen nature into the world of plants?"

Address

June 11, 1924

Dr. Steiner expresses satisfaction that this group of farmers and scientists will experiment with the guidelines given in these lectures. "They can only be guiding lines" implies a great deal of room and a rejection of dogmatic interpretation. He then offers previous experiences that indicate presenting his ideas to the general public should be reserved until the practical application of the guidelines are confirmed. He compares the details given in these lectures to the letters of the alphabet. This information from the lectures has to be assembled into meaningful units, just as words are composed from letters.

Farmers and scientists with decades of a different understanding will not change their views immediately, but momentary success is not what matters. What matters is the continuous working with iron perseverance.

Farmers need to protect their own skin, as they alone understand their farms. The scientists need to hear from the farmers how the soil and what's above it are working together. This includes the nature of the soil, kinds of woodland, what's been grown and what the yields are. These are the things farmers must know to run their farms in an intelligent way, with "peasant wit."

"I grew up entirely out of the peasant folk, and in my spirit have always remained there," Steiner says, explaining his loving devotion to farming. As a youth he planted potatoes, bred pigs and lent a hand with the cattle. Remembering his small peasant farm upbringing, he suggests the most valuable farmer is the small peasant farmer who himself as a little boy worked on the farm.

"I have always considered what peasants and farmers thought about their things far wiser than what scientists were thinking." He hopes these lectures carry some genuine "peasant wit" into the methods of science. Working as a little boy on a farm, "will serve me far more than anything I have subsequently undertaken," which included extensive academic and scientific training.

Lecture Four

June 12, 1924

Steiner begins lecture four with the observation that the world cannot be judged from conclusions drawn by investigating tiny, restricted spheres. For example, science had recently corrected itself in regard to daily protein consumption, from over four ounces a day to under two. It is understandable that a science that only recognizes coarse material forces and substances will have to keep correcting itself.

This is notable in how science views soil fertility and human nutrition by focusing just on the nutrients and ignoring the dynamics. The greater part of what plants and humans ingest is there to give them living forces. These forces help us receive substances from the very small, diluted amounts in the air through our skin and breath.

To help gain insight into the working of substances and forces, let's consider a tree and compare it to a mound of humus rich earth. Mounding up earth gives it more vitality and makes it easier to permeate with humus, much like we do in making compost piles. This earthly matter, permeated by humus, is on its way to becoming a plant, but as a mound it doesn't go as far as the tree.

Life continues from plant roots into the surrounding soil by the underground network of fungal hyphae, whose length boggles the imagination. The reason for fertilizing is to enliven the earth so this underground microbial network can help the plant get what it needs. We must know this in order to gain a personal relationship with our farming work, especially when working with manure and compost.

We can assess the vitality of living organisms with our sense of smell. Any living thing, from a cell to a whale, always has an inner and outer side, separated by some kind of skin. The insides will concentrate and organize the smells, while the outside lets them go. Plants tend to absorb odors. Steiner then enthusiastically tells us to perceive the helpful effects of a fragrant meadow of aromatic plants.

Manuring is essentially communicating livingness to the soil, yet not only livingness. We must also enable the nitrogen to spread out into the soil. With its attraction for carbon, it can carry the life property to the plant's roots. Lime helps with this, too. Because water soluble fertilizers lack carbon, they can't bring life to the earth. Although they can stimulate plant growth, they are harmful to the soil's life.

In compost we have a means of kindling life in the earth itself. Not only does compost have the living oxygen principle, but the carbon materials absorb the sentient nitrogen principle, and this is most important. Manures already have the digestive, alkaline lime principle working well enough. But for composts without manure, a small amount of quicklime, a couple pounds per ton, will help absorb the oxygen without volatilizing the nitrogen as ammonia, which too much lime will do if added to manure. The lime absorbs not only the nitrogen but also the oxygen, so fertilizing with this compost communicates the nitrogen directly to the earth without going through the soluble nitrate phase. We'll get good fodder and hay when we fertilize meadows and pastures with such compost.

Keep in mind that our detailed measures must still depend on our inner feelings, which will develop once we perceive the whole nature of the process.

We can also develop the necessary personal relationship with our sense of smell. We want the compost heap to smell as little as possible, so we pile it up in thin layers, covering the layers with peat, old hay, rotten wood chips and soil. The nitrogen, which otherwise might evaporate, is now held by the carbon in these materials.

Organic entities, like a compost pile or a cow, have streams of forces pouring both outward and inward. Horns and hooves of cows don't allow the cow's internal forces out. Instead the horns and hooves ray them back into the cow's digestive tract. The horn is especially well adapted to ray back the living forces of oxygen and the sentient forces of nitrogen.

Antlers are altogether different and discharge streams and currents of forces outward. Animals with antlers are nervous and quick, while horned animals such as cows are calmer and slower.

The grass a cow eats is not completely excreted for many days. During this time a host of microbial activities take place in her belly, and the manure is permeated by forces that carry nitrogen and oxygen. Forces in the digestive tract are like those in a plant. Manure has a life giving and sentient influence upon the soil, and upon the earthly element, not just the watery element like artificial fertilizers. The increase of microbes indicates the manure is in good condition to support life, but inoculating it with bacteria is not how to improve it.

Instead, take cow manure, stuff it into a cow's horn and bury it in good soil in the fall after the equinox. This preserves the oxygen and nitrogen forces of the manure just as the horn did this for the cow. The manure inside the horn is enlivened with these life and sentient forces which attract and concentrate the plant-like living oxygen and animal-like sensitive nitrogen forces within the soil. The manure in the horn transforms into a humus rich fertilizing substance with "a highly concentrated, life giving manuring force."

Life forces beget more life, as organization flows from lower to higher concentration. Creating something rich in life and intelligence breeds more of these qualities. Throughout the winter the content of the horn becomes inwardly alive, because winter is when the earth already is most inwardly alive. By contrast, the life of the earth flows outward in summer into the growth and activity of all the plants and animals.

The resulting "horn manure," dug up at the end of May, is stirred quickly in a pail of slightly warmed water, changing the direction after each deep vortex is formed. As it seethes around in the opposite direction, bubbles show that air is being incorporated. After an hour of stirring in alternating directions, a thorough penetration is achieved and there arises a delicately sustained aroma.

This personal relationship is extraordinarily beneficial, and you can easily develop it. The contents of the pail are then sprinkled over the land. Combine this with ordinary use of compost and manure and you will soon see how great fertility can result. These measures lend themselves to further development and can be followed up at once by another.

Grind quartz crystals into a fine, floury powder and add water to make a mush the consistency of a very thin dough. Fill a cow's horn with this, bury it over the summer in the soil, and dig it out in late autumn.

In this case you need far less quantity, maybe just a fragment the size of a pea. You have to stir this, like the other, for an hour. You can use it to sprinkle plants externally and it will prove most beneficial with vegetables.

You will soon see how the cow horn manure drives from below upward, while the cow horn silica draws from above.

Agricultural research investigates ways to increase production for financial profit, but the most important point is not quantity but quality. If you pursue agriculture in the way Steiner describes, the result can be no other than the very best food for humans and animals. "The end in view is the best possible sustenance of human nature."

A living thing has streams of forces going outward and currents of forces going back inward from the skin, and is surrounded by many streams of forces

Discussion

Question and answer sessions took place after lectures four, five, six and eight. I have included a few excerpts from these discussions.

June 12, 1924

Steiner is asked whether one can use a mechanical stirrer instead of stirring by hand, and he replies there is a difference. When you stir manually, all the delicate movements of your hand will come into the stirring. Even the feelings you have may then come into it. He then compares this to how the enthusiasm of a doctor can have a strong effect on the remedies he gives his patients. Very much can come from the hand. With enthusiasm, great effects can be called forth. As a result of the concentration and subsequent dilutions, it is no longer the substances as such, but the dynamic, radiant activity that is working.

It is best to get fresh horns from medium aged cows that have been living near your farm for three or more years. They should be 12 to 16 inches long and can be used for three or four years. Don't disinfect the horns. People go too far in disinfecting things.

To store the horn manure, make a box upholstered with a cushion of peat moss on all sides so the strong inner concentration will be preserved. But once stirred in water the horn manure should be used right away. The microbes inherent in the horn manure are propagated during the incorporation of air and water, but cannot live long after the stirring stops. The horn silica can be stored where the sun can shine on it. This will do it good.

To the question, "With the fine spraying of the liquid due to the spraying machine, will forces be weakened?" Steiner replied, "Certainly not; they are intensely bound. Altogether, when you are dealing with spiritual things—unless you drive them away yourself from the outset—you need not fear that they will run away from you nearly as much as with material things."

"What about using machines on the farm?" is the next question. Steiner responds by noting, "You can hardly be a farmer nowadays without using machines." As a farmer though, you need not be crazy for machines.

To questions regarding stronger concentrations for more rampant growth, Steiner recommends staying with the amounts given. This is now accepted as 1/3 cup horn manure in three gallons of water for one acre, and 1/2 teaspoon of horn silica in three gallons of water for one acre. The former is applied in the evening before sowing so as to influence the soil itself, the latter is applied in the morning on the growing plants.

"Can anyone you choose do the work?" is a broad question. Some people have a "green thumb" and others don't. As described in lecture three, such things will come about simply as a result of the human being practicing meditation. "When you meditate you live quite differently with the nitrogen which contains the imaginations. You thereby put yourself in a position which will enable all these things to be effective."

There were times in the past when people knew that by certain definite practices they could make themselves fit to tend the growth of plants. Delicate and subtle influences are lost when you are constantly living and moving among people who take no notice of such things. When talking about practicing certain concentrations to combat parasites, Steiner suggests that you would want to choose the proper season. Establish a festival between the middle of January and the middle of February when the earth unfolds the greatest forces.

Old folk proverbs contained a peasant's philosophy, a subtle wisdom about such timing. One marvels to see how much the peasant knows of what is going on in nature. Steiner regrets the loss of the cultural philosophy common in his youth, when "you could learn far more from the peasants than in the university."

Someone asked about determining the number of cow horns purely by feeling. Steiner doesn't advise this, saying we must be sensible. Test it thoroughly according to your feeling, but then translate the results into figures. Bear in mind that the judgments of the world are tending toward calculable amounts, so compromise in this respect as much as possible.

Sandy soils will need less lime and clay soils will need more. Clay soils are often oxygen deficient.

A layer of peat or earth put all around the compost pile, after it has been turned, protects it.

Kali magnesia, known as sul-po-mag today, is the kind of potash Steiner meant.

The last paragraph in the discussion is extremely important. Steiner points to relearning how farmers fertilized before artificial fertilizers were available. "You must remember that the cow horn manuring is not intended as a complete substitute for ordinary manuring. You should go on manuring as before. The new method should be regarded as a kind of extra, largely enhancing the effect of the manure hitherto applied. This latter should continue as before." (See Appendix, Lessons From Old Agricultural Textbooks)

Humus-substances in the process of decomposition on the way to becoming a plant

Lecture Five

June 13, 1924

The agriculture course has 174 pages, so at page 87 we are exactly halfway through it. Steiner now reiterates the last few sentences, "The preparation I indicated yesterday for the improvement of manure was intended, of course, simply as an improvement, as an enhancement. Needless to say, you will go on manuring as before."

A thorough study of how farming was conducted in the latter half of the 19th century sheds light on the phrase "manuring as before." All farms had livestock and were encouraged to make compost and carefully store and use their manure. Crops included legumes and followed strict rotations around the farm. Green manure cover crops and the liberal use of lime were considered necessary. In a market garden, it was common to spread 40 to 50 tons of composted manure per acre annually, and upwards of 100 tons was not uncommon. Soil moisture and microbes were conserved by careful attention to tillage, and plowing was considered an art.

It is of great value to recognize that the soil is a continuation of the plants growing in it. Science no longer perceives this common life of the soil and all plant growth, nor how it continues in the manure. The forces taken from crops on our farms must be restored. The manure must be subjected to proper treatment to vitalize it, but simply inoculating it with bacteria is not sufficient. That would be like adding flies to a dirty room to clean it, rather than working on the room itself. Adding artificial fertilizers won't help either, at least not for long. Because of their water solubility, they cannot fertilize the earth directly. We can only do this by working with solid organic matter.

Steiner disagrees with science's insistence that the elements found in plants are the only essential ones, and should be added in a water soluble form. This ignores the role of silica, lead, arsenic, mercury and sodium, which are regarded as mere stimulants. But they are the most important ones of all, and are provided freely with the rain. We don't have to take them into account like the potash, phosphate and limestone, which we must fertilize and till the soil for.

Human activities can utterly prevent the soil from receiving silica and the rest. They become unavailable for plant growth when soils are compacted, or when we use artificial fertilizers and till randomly and carelessly. When this is the case, subjecting the manure to further treatment will help. It is not substances but living forces which should be added. It will be easy to use minute quantities in the right way to set free radiant forces in manure and compost.

Steiner then mentioned, as a general indication, preparations to add in minute doses to the manure and compost to bring the nitrogen nature into the world of plants. Replacements can be found if certain things are difficult to find. We must provide the right way for carbon, oxygen, nitrogen, hydrogen and sulfur to come together with other substances, notably with potash. It is very important that the potash assimilate itself rightly with the protein substance in the plant.

To do this we take yarrow[6], a miraculous creation as a model plant that shows how to bring sulfur in a right relation to the substances of the plant, particularly the potash. We clip off the florets. If you have the dried flowers, moisten them with yarrow tea. Sew them up in the bladder of a stag. Hang it up in a sunny place during summer and in the fall bury it shallowly in good soil until spring.

Put a few grams in a manure or compost pile, which can be the size of a house, and the radiant power will influence the whole mass. This endows the pile with the power to enliven the soil so that it receives silica, lead, arsenic, mercury and sodium from the atmosphere. The minute amounts of sulfur in yarrow, combined in a model way with potash, enables the yarrow to ray out its influences through large masses and great distances.

A stag has antlers and is intimately aware of the periphery in its environment and to what is going on all round it. In the stag's bladder we have the necessary forces to give yarrow the power to enhance its own ability to combine sulfur with the other substances. It is important to observe that we have an absolutely fundamental way of improving manure while remaining within the realm of life and not going into inorganic chemistry.

Next is another example. We want to give the manure the power to receive so much life into itself that it transmits life to the soil. We must also humify the manure so it can bind together the substances necessary for plant growth. In addition to potash we need calcium compounds. While yarrow assimilates potash with the help of sulfur, another plant also assimilates calcium. That plant is chamomile[7]. You can look around and trace the process which

---

[6] *Achillea millefolium*

[7] *Matricaria recutita, or German Chamomile*

these herbs undergo when taken medicinally. In yarrow we have an effect on the bladder and kidneys, but chamomile is used for an effect on our intestines, for a stomach ache and to help our digestion.

So we clip off the beautiful, delicate yellow and white flower heads of chamomile and stuff them into cow intestines. Take these precious little sausages and bury them shallowly in humus rich soil. Choose a spot where snow will remain and leave them through the winter. Add a small amount to the manure, as you did with the yarrow preparation, and you will get manure with a more stable nitrogen content. It will kindle life in the soil and will create much more healthy plants if you fertilize this way.

Stinging nettle[8] is the greatest benefactor for plant growth in general, and it would be hard to find a replacement. Besides sulfur, stinging nettle carries potassium and calcium in its radiations, and also iron. These iron radiations are almost as beneficial to nature as they are in our blood.

To improve the manure still more, bury a mass of the leaves and stems of stinging nettle, surrounded by peat moss, in the soil. Be sure to mark the spot as it is easy to lose the place. Dig it up after it has spent the winter and the following summer underground. Use small amounts like the other preparations and the manure becomes inwardly sensitive, aware and we might say intelligent. It won't suffer from loss of nitrogen. "It is like a permeation of the soil with reason and intelligence."

The prevention of diseases takes a more general course in plants than in animals. A rational improvement of manure, using the following method, can remove a large number of plant diseases. The

[8] *Urtica dioica*

manure must bring calcium into the soil, but as this calcium must remain within life ordinary lime is of no use. The bark of a white oak[9] tree contains plenty of calcium in an absolutely ideal form. The outer bark is an intermediate product between plant and humus, as described in lecture four when comparing a tree to a mound filled with humus.

This calcium has the property of damping down life and restoring order when growth gets rampant. Calcium's unique structure in oak bark helps this process without creating shocks in the plant. We grind the bark up and pack it into the brain cavity of a cow or other domestic animal's skull and cover the spinal cord hole with a bone. This time we bury it in a marshy damp place with decaying vegetation throughout fall and winter. When added to the manure pile it lends forces that arrest and reverse nitrification and combat plant diseases.

Although these preparations require a certain amount of work, it is a lot less work and cost than the production of artificial fertilizers and agricultural chemicals. The latter will produce only superficial effects, such as a large size, but the plants will no longer have the proper nutritive power and it is inevitable that human health will eventually suffer.

Next, we need something to attract silica from the whole, distant environment. We don't notice that the soil gradually loses silica in the course of time and of silica's great significance for the growth of plants. Scientists in Steiner's time had just started accepting the idea of the transmutation of elements, in regards to radioactivity. If the processes taking place around us all the time were not so utterly unknown, the things Steiner explains would be more believable. A

---

[9] *Quercus alba*

modern agricultural student might now argue that Steiner hasn't told us how to improve the nitrogen content of the manure. Though it is quite a foreign concept, the improvement in nitrogen content occurs due to the alchemy in the use of yarrow, chamomile and stinging nettle.

If potash is working properly through herbs, animal organs, and the resulting biology, this hidden alchemy transmutes the potash into nitrogen. It even transforms the limestone into nitrogen. Soils with active calcium grow legumes, whose symbiotic bacteria incorporate atmospheric nitrogen into the soil biology. This transmutation requires microscopic organisms in the soil.

The ratio of nitrogen and oxygen in the air is approximately four to one, which is similar to the relationship between lime and hydrogen in the organic process. Steiner maintains that under the influences of hydrogen and carbon, lime and potash are constantly being transmuted at length into the nitrogen which is of the greatest benefit for growing plants.

The elements calcium and potash become part of the bodies of microbes. As they work their way up the food chain, nitrogen from the air in the soil also becomes part of these microbes. As Steiner points out, these processes are not completely understood and can therefore be called mysteries. After minerals are incorporated into a microbes's body, nitrogen gets coaxed into the body with the help of carbon and oxygen. It then becomes part of amino acids, the ideal form of nitrogen for plant growth. All this happens under the influences of hydrogen, or you could say under the influences of the entire heavens, because hydrogen is by far the most common element in the universe.

Silicon is also transmuted in living organisms, such as the forming and dissolving of an insect's exoskeleton. In plants there must arise a clear and visible interaction between the silica and the potassium, but not the calcium. Lime and silica are polar opposites. We can fertilize the soil so it aids in this relationship by finding a plant whose own relationship between silica and potassium can impart this power to the manure.

Steiner now introduces dandelion,[10] a kind of messenger from heaven who mediates between the tiny amounts of silica, distributed in the atmosphere and beyond, with what is needed in the soil. Once again, we must expose the dandelion to influences in the soil over winter. After gathering them and letting them wilt a little, we are advised to sew them up, pressed together, in a cow mesentery. Then we bury these pillows, and by spring they will be thoroughly saturated with these distant influences.

Add this to manure or compost and it will give the soil the ability to attract just as much silica from the atmosphere as the plants need to make them sensitive to what is at work in their environment. Plants grow better if they can sense all that is in the soil and above them, and enlist these things to help them grow. They do this with the aid of their symbiotic microbial network.

Although you can artificially fertilize and limit the environment the plant needs, it is not good to do so. Treat the soil with these preparations and the plant can draw what it needs from a wide circumference, including adjacent meadows and forests. Steiner proposes we would create better fertilizers by adding yarrow, chamomile, stinging nettle, white oak bark and dandelion, or suitable substitutes, to our compost and manure instead of using all manners of chemicals.

Finally, before using the compost or manure, press out the flowers of valerian,[11] dilute the extract, and stir briefly as with the horn manure. By sprinkling this over the compost or manure you

---

[10] *Taraxacum officinale*

will stimulate it to behave in the right way to phosphorus. Phosphorus, which tends to lock up, can be made available by the biological activity the valerian flower juice stimulates. With the help of these six ingredients excellent fertilizers can be produced, whether from liquid manures, farmyard manures, or composted organic materials.

[11] *Valerian officinale*

Discussion

June 13, 1924

In response to a question, Steiner replies that manure does not need a roof over it. Although too much rain would wash it out, in moderation rain water is good for the manure. A skin of hay or peat is also a good idea. In response to a question regarding a new method of loosely piling manure to generate warmth and become odorless, Steiner reminds us that success with new methods may not always be practical in the long term. A new medicine can work wonders the first time you take it, but its curative effects may diminish if you keep taking it.

The spontaneous generation of warmth in manure piles is exceedingly good for the manure, and becoming odorless indicates the method is beneficial. The pile does not need to be in a pit or on a hill, just build it up from the normal ground level. Don't pave underneath it, but add clay if the soil is sandy or sand if the soil is clayey.

One of the farmers present had been eradicating yarrow and dandelion, thinking they were bad for cattle. Steiner suggests watching it carefully, that an animal will not eat what is not good for it. We only need a small amount of the herbs he is recommending. When we want to stimulate something alive, we often use what we would not otherwise use. After all, medicines are generally poisonous. The process, not the substance, is what is important.

It doesn't matter how long the preparations are kept with the manure, but they should be used before the manure is spread on the fields. The preparations are not buried together, but in separate spots on the farm. Animals, from earthworms to dogs, can dig them up, so enclosing the preparations in a nylon screen and covering the spot with big rocks is recommended. It can be overgrown with plants. In the pile, place them separately in holes 12 to 18 inches deep and close the manure up around them. It all depends on radiations, so not placing them too near the surface or each other will keep the radiations from going outside or interfering with one another.

The outermost bark of a white oak is taken from a live tree. The other preparations are buried in fertile soil, not the subsoil, and frost will do no harm. The earth, by virtue of the frost, is most intensely exposed to distant influences.

The silica quartz or feldspar crystals are ground down in an iron mortar and pestle. An iron, T-post tamper works well. It can be further ground between two plates of glass. You will want to wear ear plugs and be careful not to breathe the dust.

In the absorption of food, the forces developed by the body are the essential thing. The animal must receive the proper food to be able to develop sufficient forces to absorb what it needs from the atmosphere. The food is there to stretch the organism to enable the animal to receive what it needs from the atmosphere. Too much food can shorten an animal's life. There is a happy medium between maximum and minimum.

Radiations of a preparation in a compost pile

Lecture Six

June 14, 1924

Now Steiner gives us examples which, taken as a starting point for experimentation, will lead us further into ideas regarding harmful plants and animal pests. For instance, a weed might be a useful plant, although growing where we don't want it. To deal with plants growing out of place we are first reminded of the distinction between the forces of growth and reproduction working downward into the soil, and the forces of nutrition working upward into the air. The former are influenced by lime, and the latter by silica.

Both the soil and the traditions have become exhausted, though sometimes simple peasant folk can lend a hand. For example, science is at a loss to find a remedy that prevents aphids on grape shoots. Aphids are beneficial in the soil with their interaction with ants and fungi, but detrimental in the plant canopy. The solution needs to recognize that there are soil forces up in the plant that should be down in the soil. People used to know these things.

Influences from the inner planets enter the soil and ray back up into growth and reproduction[12]. On the other hand, all that makes a plant fill out and become nourishing comes from the forces of the distant planets. This is the insight that gives us an idea on ways to affect plant growth.

Weeds are often medicinal plants, influenced by the moon. We can see sunlight reflected by the moon, but not the other forces. Lunar forces strengthen the soil, intensifying its normal vitality to the point where plants can reproduce themselves. Reproduction is simply an enhanced growth process, and lunar forces are what enhance plant growth to the point of reproduction.

Sowing by the moon signs was part of the old instinctive science. But nature doesn't punish us if we are inattentive to moon cycles when planting or harvesting. If we sow, till or fertilize at the wrong time, for example not at a full moon, what we've done will simply wait in the soil until the next full moon.

Weeds become particularly troublesome in wet years when you can't get in the fields to cultivate. However, we can learn how to weaken lunar influences and make weeds reluctant to grow. Gather the weed seeds and burn them in a wood fire. The ash concentrates the opposite force to what water attracts as lunar forces. By scattering this small amount of ash over the fields, the weeds will grow less rampantly. We may have to repeat this every year for four years, as these influences occur in four year cycles.

People used to know these things instinctively and could have the plants they wanted wherever needed. Steiner says that he can

---

[12] *Confusion has risen with the contradiction of the direction of forces here compared to lectures one and two. Some postulate secondary forces going in opposite directions, and others wonder if either Steiner or the stenographer got mixed up.*

only give indications, but they are quite practical. As far as proof goes, if he had a farm he would apply the methods at once. Tests can verify it, but can also be misleading. But one can know things inwardly, and what is inherently true will later be confirmed.

We will have to be less general when dealing with animal pests than with weeds. Let's take the field mouse as an example for our experiments. Poisons of phosphorus and strychnine were used to kill mice. State regulations were made requiring neighborhood farms to do this too, or the mice just come back from there.

What we do would also be better if the neighbors followed suit, but a general insight into the method is required for neighbors to see the sense in doing it. Intelligent insight works much better than police regulations.

Catch a fairly young mouse, skin it, and burn the skin when Venus is in the constellation of Scorpio. When we move from plants to animals we need ideas that consider not just our solar system but the fixed stars, notably the constellations of the zodiac. The word zodiac means "animal circle." Although lunar influences are enough to bring reproductive forces to plants, animals also need forces from Venus for reproduction.

As with weeds, the reproductive force is weakened by fire. The burned mouse skin inhibits field mouse reproduction. Sprinkle this ash over the fields and the mice will avoid the area. Again, only a small amount is needed.

We can get much pleasure in farming this way, reckoning with the influences of the stars without becoming superstitious. However, to deal with insect pests, we need to understand that they are subject to different starry influences than the higher animals.

For example, a nematode infesting sugar beets makes the roots develop swellings and the leaves look limp in the morning. Steiner reminds us again that leaves absorb atmospheric influences and that roots absorb influences from the soil. Nematodes appear when the leaf and air forces are pressed downward into the soil. All living creatures can only live within certain limits and with certain forces. Aphids can only live above ground when forces normally in the soil are up in the plant's leaves. Root knot nematodes live when forces normally in the air are found in the soil.

Burn the entire insect when the sun is in the constellation Taurus, which is the opposite sign from Scorpio. The sun is different in different signs, shining down on earth at different angles. Spread the ash on the infested fields and the pest will shun life there, especially if you do this each year for four years.

It is important to relate to the soil this way. Just as water brings about fertility, fire destroys it. Burning seeds annihilates their power of fertility. Burning insects and animal skins in the appropriate constellations annihilates the power of fertility in these animals.

To deal with plant diseases we need to notice these arise when the forces of life are not strong enough to mitigate the effects of overly strong lunar forces. This happens especially when a wet winter is followed by a wet spring. The earth becomes too strongly alive and pushes growth too fast. Then the above ground part of the plant becomes a lush medium for parasitic growths such as mildews, blights, and rusts. The overly intense lunar influence from excessive moisture interferes with the forces of forming and ripening seeds, which come from the outer planets.

Several significant relationships are involved. Wet conditions in the soil overwhelm the lunar growth forces, producing nitrates in the wet soil. This results in plant diseases. When the soil is dry the life-giving influences of the moon are not as strong. Although the lunar forces are necessary for seed formation, they must not be too strong.

We can relieve the soil of excessive lunar forces by perceiving what works in the soil to prevent it from absorbing the excessive lunar influences in the water. What counters this lunar activity is the outer planets and silica. Horsetail[13] is a plant rich in silica that we can use in small quantities. Prepare a decoction of dried horsetail by simmering it for 30 minutes. Dilute this and sprinkle it over the fields. Understanding the influence horsetail has on our kidneys will give you your guiding line. This is not so much a healing process as it is the opposite process to the overly intense lunar forces. Silica is the antidote for too much water, as it absorbs excess moisture.

Real science arises when we learn how to control the overall forces at work. Life cannot be understood in isolation from the whole, just as we can't understand why a compass needle points north without considering the whole earth with its magnetic field. Modern science's analytical, microscopic way of studying nature must give way to an understanding of the macroscopic, entire universe from which life proceeds. Steiner tells us, "Nature is a great totality; forces are working from everywhere."

---

[13] *Equisitum arvense*

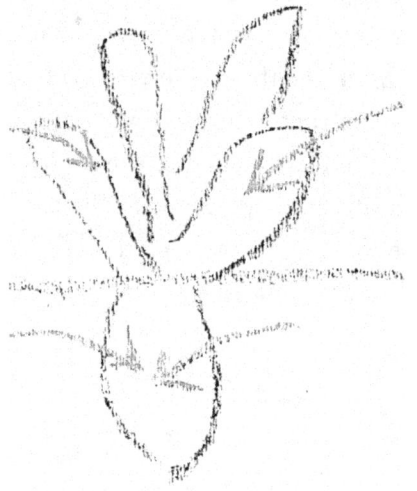

In the nematode-infested plant, the forces that should be working up above are working down below

Discussion

June 14, 1924

For the insect "pepper" you can use both the larvae and the adult, although it might shift the constellation. The proper constellation moves from Aquarius to Cancer as you pass from the winged insect to the larvae. For a mature insect the proper constellation might be more towards Aquarius. It will have to be tested to see how different kinds of lower animals respond to ashing with the sun in different constellations.

For animals, Venus should be behind the sun and in front of Scorpio. The ash is sprinkled on the field like you would put pepper on your food. It has a great radius of influence, and it can also be applied to fruit trees.

The next question was whether the specially prepared manure, besides being given to turnips and garden crops, is also important for grains? Steiner said to continue the methods that have proven good, and supplement them with what has been given here. The influences will be less with sheep or pig manure than with cattle manure, but go on manuring as before.

To a question about inorganic fertilizers, Steiner responds that mineral manuring, as with NPK fertilizers, must cease because the products from fields thus treated lose their nutritional value. "It's an absolute general law," he says. With the methods he has given we'll be able to gradually stop using chemical fertilizers. These methods will be much cheaper and the soluble fertilizers will go out of use. Steiner suggests we consider how everything is being mechanized and mineralized, yet minerals only work as they do in nature. Do not put lifeless artificial fertilizer into a living soil.

In a recent discussion on beekeeping, a modern beekeeper was especially keen on the commercial breeding of queens. Queens are sold in all directions nowadays, instead of merely being bred within the single hives. Steiner had to reply that modern opinions are based on far too short a period of time. "No doubt you are right; but you will see with painful certainty, if not in 30 or 40, then certainly in 40 to 50 years' time, that beekeeping will thereby have been ruined."

Lecture Seven

June 15, 1924

Although things in nature are frequently studied by themselves, Steiner begins the seventh lecture by emphasizing that all things in nature are in mutual interaction. In times past farmers were thoroughly familiar with the interactions of minerals, plants, and animals through an instinctive insight which has now been lost. Today, scientists only study the coarse interactions but not the effects of the finer, more subtle ones.

We must observe these more intimate relationships, which are constantly taking place, when we are dealing with the life of plants and animals together on the farm. Besides our crops and livestock, we must observe with intelligence the colorful world of insects, hovering around plants, and learn to look with understanding at the birds. Modern society has caused a reduction of bird populations in certain areas, and doesn't realize how greatly farming and forestry are affected.

When we look at and consider a tree, it's noticeably quite different than an herbaceous plant. We can only include the flowers, leaves and shoots growing out of the tree branches annually as the plant nature. The rest of the tree, the older branches and trunk, is more like a mound of soil which the plant nature is growing upon. Right underneath the bark is the cambium layer that connects the new growth to the soil, like a long thin root for those "plants" growing on the branches.

To help us understand what a root actually is, Steiner makes this comparison. If a bunch of plants are growing close together, their roots would intertwine and merge with one another. As you can imagine, such a complex of roots would not remain a mere tangle of one root winding around the other. "It would grow organized into a single entity. Down there in the soil the saps and fluids would flow into one another. We could not distinguish where the several roots began or ended. A common root being would arise for these plants."

This is the role of fungi, whose hyphal systems join the plant roots together. The common mycorrhizal network is what these connecting fungal roots are now called. Communication happens between plants through this network. A plant infested with aphids, for example, will within a day transfer a signal that elicits a nearby plant to release odors which attract aphid eating insects, thus protecting itself. There is no hard and fast line between the life in a plant root and the life in the soil because of this common mycorrhizal network.

The cambium layer is where cells divide and make new cells. This is the living, growing layer underneath the bark. The tree's trunk is like soil that has bulged upward into the air. Having grown outward, the tree needs more inwardness, more intensity of life than an herbaceous plant does.

Trees grow high up in the air where the atmosphere is different than directly over the soil where other plants grow. There is more of the oxygen process happening around herbaceous plants and more of the nitrogen process up in the treetops. A deep dark forest feels different than a sunny meadow.

Steiner maintains we can easily become clair-sentient with respect to the sense of smell, especially if we acquire a certain sensitivity to the aromas of herbaceous plants compared to trees. The scents wafting from blooming fruit trees and forests are rich in nitrogen, while the other plants smell more earthy. Steiner recommends you "accustom yourself to specialize your sense of smell—to distinguish, to differentiate, to individualize, as between the scent of earthly plants and the scent of trees."

Because of this difference, which we can learn to smell, the life and oxygen forces are depleted underneath where trees are growing. Tree roots last a long time and become more mineralized. Clearly envisioning this, we look about us and see insects living and moving in the air, while down below the insect larva live with the roots and soil.

Also in the soil are the wonderful earthworms. Study how they live together with the soil. Their castings are a perfect example of a humus substance containing what the plant root needs in an available but colloidal form. This allows plants access to nutrients, but since they are not in a water soluble form they don't leach out when it rains. Farmers must take special care of the soil to encourage the extremely beneficial activity of earthworms.

There is a remote similarity between insects and birds. Insects flutter about primarily in the meadows and leave the treetops for the birds to fly around in. Both help to distribute nitrogen forces wherever they are needed. Winged animals are unthinkable without plants and vice-versa. Farmers should have some understanding of the care of birds and insects, for in nature everything is connected.

Steiner says, "These things are most important for a true insight, therefore let us place them before our souls most clearly." What forests do for the land far around has to be done by quite a few other things in unwooded areas. The growth of soil is subject to other laws in areas where forest, field and meadow alternate, then in wide stretches of unwooded country. Observe and then contemplate these things clearly.

We should have the insight to preserve forests in districts where they were before human intervention. Forests are good for the surrounding farmland. We should have the heart to increase woods when we see vegetation stunted, or make clearings in the forest if we notice plants growing rampant and not producing seeds. The regulation of woods and forest is an essential part of agriculture.

The world of earthworms and other soil life is related to limestone, the mineral nature, and the world of insects and birds is related to silica. Things like this were recognized in olden times by instinct. Earthworms require lime in the soil, and birds need some silica rich conifers around them. Realizing this, another kinship emerges, the inner kinship of mammals to shrubs and bushes. It helps to plant shrubs in our landscape because mammals love them.

Another intimate relationship is the one of mushrooms or fungi to bacteria and harmful parasites. The meadows on our farms should be well-planted with mushrooms and toadstools. A meadow rich in beneficial fungal activity helps prevent harmful fungi and bacteria from affecting the plants we grow on the rest of our farm.

"So, we must look for a due distribution of wood and forest, orchard and shrubbery, and meadow lands with their natural growth of mushrooms."

It is not economical to rid our farms of these things, hoping to increase crops. We will get worse quality by increasing tilled acreage at the cost of removing these other things. Farming is intimately connected with nature, and to engage in it you must have insight into these mutual relationships of nature's husbandry.

Next Steiner asks, "What is an animal? What is the world of plants?" We will have to ponder these questions, and the relationship of plants and animals, in order to learn how to feed our livestock. By observing what is in a mammal's environment we can perceive what is happening.

A mammal receives and assimilates in its nervous system, and part of its breathing system, air and warmth. It is a creature that lives directly in the air, breathing in oxygen, and creating warmth in its blood flow. But animals cannot assimilate water and earth directly. They must drink and eat, so they must have a digestive system to do this.

On the other hand, a plant doesn't need a digestive system. It has an immediate relation to water and earth, just as the animal has with air and warmth. The plant lives directly with water and soil, but the converse is not true. Plants do not assimilate air and warmth internally like an animal does with water and earth. Plants give off oxygen and a slight degree of warmth.

The plant is inverse to the animal. The expelling of air and warmth has the same importance for the plant as drinking and eating does for animal. The plant lives by giving. Everything in nature lives by give and take. The symbiotic relationship between plants and animals is harmoniously regulated by their environment.

Orchards, shrubs and forests are regulators to give the right form and development for the growth of plants throughout the whole farm. Earthworms and lower animals, in unison with the soil's lime content, act as regulators down in the soil. That is how we must regard the relation of our tilled fields to the orchards and the forests, with their insects and birds, and to the mushroom rich meadows with the grazing livestock. "This is the very essence of good farming."

A common root being showing the underground mycorrhizal network

Lecture Eight

June 16, 1924

In this final lecture Steiner wants us to develop insight to act individually with intelligence on the practical hints he still has to offer.

Plants have their physical mineral body and also their living body that builds and maintains their physical presence. Hovering around them is an awareness and intelligence that connects with the plant when it produces edible nourishment in support of animal and human awareness and intelligence.

Insight reveals whether or not something supports some process in animals or humans. These days there is much confusion concerning animal as well as human nutrition, which shows up in such things as imagining food is ingested and burned or combusted in the body. Yet, combustion is a process in lifeless nature that is quite different from what takes place within the body where the process is altogether living and intelligent. To understand what goes on we can observe animals.

Our human system is threefold, consisting of the head pole, with limbs and metabolism on the other pole, and a well-defined rhythmical system in between, comprised of the heart and lungs. Animals have the two outer poles, but the middle, rhythmic one is not as independent as it is in humans.

The substances of the head, nerves and sense organs come from the soil through what is digested. The substances that make up the limbs and metabolism, on the other hand, come from what is absorbed out of the air and warmth above the soil. This is important.

The opposite is true of the forces. In the head, with its nerve and sense system, are forces of a distant, universal origin. The rest of the body is subject to local, earthly forces, such as gravity. In practice, you need to feed a work animal so that it can absorb the substances out of the air to strengthen its body. What is needed by the head, rather than the muscles, must be gotten from the actual fodder.

Distant and atmospheric substance cannot flow easily into a dark stable. The animals should be out on pastures and have the opportunity to sense the surrounding world. Compare the confined animal, eating what humans decide it needs, to one that is using its own sense of smell to find food outside. The indoor animal has inherited traits that conceal its lack of forces from outside, but its descendants will show the lack. Thus the next generation of animals will become weak and won't be able to absorb what they need to be truly healthy.

In the head you have the brain, which in humans serves as the basis for our ego, our self-consciousness. The animal does not have self-consciousness. Its brain is only on the way to forming an ego, but it is not there. The substance of the brain comes from what the animal has eaten. Some of the earthly substance gets digested and excreted, but some continues on to finally be deposited in the brain.

In humans, more of what's in the belly gets to the brain, in the animal it is less. So, the animal's manure has more potential for self-consciousness, whereas in humans that potentiality proceeds to the brain. Animal manure brings the potential for self-consciousness to the roots of plants. This potentiality connects the roots to the soil's microbes and minerals, so these plants can access all they need and in turn provide the best possible nourishment.

A farm is a living organism, with nitrogen forces developed in the fruit trees and forests, and oxygen forces in the meadows. As its animals eat the pasture, they develop forces which are given over to their manure. These forces enter the soil with the manure and cause the plant roots to grow deep in the earth, following gravity.

A farm is also an individuality, and you will gain the insight that your animals and plants should be kept within this mutual interplay. You hinder the way nature works by bringing in nitrogen fertilizers. It is a perfect, self-contained cycle to let the farm's own animals fertilize the farm. Each farm, as a unique individual entity, requires the right kinds and amounts of animals.

All we can do here is indicate general guiding lines. As far as possible we should make our farms able to sustain themselves without becoming fanatical about it. In life, with the way economics works, this may not be fully attainable, but a self-sufficient farm is something to aim for.

Observe a root and see how it absorbs the manure's forces. The plant is assisted by the forces in the manure as well as the salts in the earth. This is what becomes the food we eat and eventually nourishes our head and nervous system. A growing animal also needs this nourishment, so we feed them roots. Carrots are traditionally fed to calves.

You also need a second kind of food to help bring the forces in the head to the rest of the body. It needs to be something that rays out in the plant. Linseed or fresh hay added to the carrots will make good feed for young cattle. Any combination of a root crop with a long, thin plant like hay will stimulate the head and assist what is needed to pass downward to fill the body. In agriculture we must always learn to look at the things themselves, and learn what happens to these things when they pass, either from the animals into the soil, or from the plant into the animal.

On the other hand, for good milk production we need to strengthen the middle part of the animal. We need the right cooperation between the forces streaming back from the head with the substances that pass forward from behind. We don't need roots in this case, or blossoms or fruit. For good milk production we want what is between these, the green vegetation.

To further stimulate the development of milk we use plants which have their flowers and fruits close to their leaves. These are the legumes, most notably clover. It takes a generation to see the results. The offspring of animals treated this way will make good milk.

We can observe that when the old instinctive wisdom was lost a few things were maintained. Certain things are still known and done in farming without really knowing why. People no longer have a comprehensive vision of the relationships of forces, and you won't get this out of textbooks which make feeding livestock complicated. Taking a more simplified approach as Steiner is indicating will lead to a more comprehensive view.

Another point worth noticing is that the fruiting process can be in other parts of the plant. Although for most plants we sow the seed of the fruit, with the potato we use their "eyes." The ability to propagate doesn't always follow the flower, it can be latent in an adventitious bud. In plants like grasses that stop short of making fruits, the fruiting process can be enhanced by the combustion process, such as drying hay in the sunshine. We heat and cook food because warmth plays a considerable part in the fruiting process and makes the food more digestible. We enhance the fruiting process of plants that quickly go to seed, such as legumes, by cooking, boiling or simmering them.

Steiner then mentions the raw food movement and vegetarianism. There are two sides to everything. Someone with a weak physical constitution will tend towards laziness on such diets. On the other hand, there is an advantage to vegetarianism and raw foods for physically strong people. They summon forces out of the food which, if cooked or overly processed, would remain latent and could lead to rheumatism, gout and diabetes. These things are unique to each individual.

To fatten animals, you feed them fruiting substances, maybe further enhanced by cooking. Also, give them food which has a fruiting process enhanced by cultivation, like domesticated turnips or beets that are growing bigger than when in the wild. Oil plants are helpful to distribute what the animal absorbs. But we also need something of the root nature to allow the earthly substance to pass upwards to the head. We have roots for the head, flowers, fruits and seeds for the metabolic and limb system, and leaves for the rhythmic system in the middle. The last thing needed for the whole animal is a small amount of salt. Small quantities fulfill their purpose if the quality is right.

For those inclined to questions regarding food, much can be learned from tomatoes. They make an organization of their own in the body. This has a beneficial effect on liver problems because of all our organs the liver works most independently. On the other hand, tomatoes should not be given to a person suffering from cancer. This is a disease that is already operating independently in the body and tomatoes would worsen this condition. When growing tomatoes, we see them happiest in a compost made of plants not fully decomposed. They develop splendidly in a compost made of the leaves and stems of the tomato plant itself. This quality of its growth is connected with its power to influence, within the body, the ability to organize independently.

In the same plant family we have the potato, with its ability to pass easily through the digestive process. In the proper amounts it makes the brain independent. However, the over consumption of potatoes is one of the factors for the rise in materialism since their introduction. Knowing these things will relate agriculture to social life, and this is infinitely important.

Steiner says he could go on giving many individual guiding lines, which are only foundations for many experiments. Wait until after the experiments are confirmed by the farmer to publish them. It makes a great difference if a farmer is speaking from direct experience, or a non-farmer is speaking theoretically.

We are now at the end of the lectures. Besides the deep inner value of what has taken place here, intended as real, useful work, our hosts have given it a truly festive setting. "Thus, with our Agriculture Conference we have also enjoyed a real farm festival."

Discussion

June 16, 1924

The astronomical indications will have to be worked out. Broadly speaking, from February to August will be good for insect preparations. For this year, 1924, the time between mid-November and mid-December would be right for the mice. The full moon is the time from a crescent moon to full, 12 to 14 days, and the new moon is from then until it disappears, another 12 to 14 days. Sprinkle the preparations soon after making them. Spread ash of horse flies over the fields, not on the horse. The ashes should be thought of as additions to the manures.

Electric currents and radiation have a negative influence on human development. Consider how electricity is now being used above the earth as radiant and as conducted electricity, to carry the news as quickly as possible from one place to another. People in the midst of electricity, notably radiant electricity, will no longer be able to understand the news which they receive so rapidly. The effect is a dampening down of their intelligence.

Salt helps food get to where it needs to go in the organism. You can use it in making silage.

Everything that enters an organism as food must undergo a complete change. Even warmth must be changed in the body. If it doesn't, we catch a cold.

The next question wonders about the karmic effects of making the inset peppers, and if it matters what frame of mind you are in. Steiner reminds us again to consider the whole way we have to think about things. In lecture eight he pointed out how one must see from the outer appearance of the linseed or the carrot what kind of process it will undergo inside the animal. "You will go through such an objective education if this knowledge becomes a reality in you at all, that it is surely unthinkable without your being permeated with a certain piety and reverence." You would have to have evil intentions to cause harm, so common morality should also be fostered. Killing a live being, like a fish, is different than destroying fish spawn. It is different if you hold back a process that has not yet been completed. Preventing the conditions for a host of beings to be born is on a different level than killing things.

A question was asked about using human feces as fertilizer. Steiner responded this should be used as little as possible, no more than what the people on the farm produce.

Green manuring has its good sides and is useful for certain plants. It has a strong effect on leaf growth.

AGRICULTURE ABRIDGED | 73

A plant is surrounded by many forces

Appendix: Lessons From Old Agricultural Textbooks

by Jeff Poppen

Lessons From Old Agricultural Textbooks

An Introduction To Soil

How Soils Form—Over time, rocks decay into soil and soil hardens into rocks. Soil is forming all the time. Long ago, revolving stars and gasses eventually condensed and became our solar system and planet with its rocks and then soil.

The forces weathering rocks into soil today are heat (volcanic activity), cold (frost and ice), wind, water and gasses (oxygen and carbon dioxide working chemically). Once formed it is called sedimentary soil. Soil is not stable but can wash downhill, called alluvial soil, or be blown about by the wind and is then called loess soil. Life depends on soil and its care is the farmer's primary concern.

Soil particles vary in size. The biggest ones are grains of sand. Sand is quartz, a combination of silicon and oxygen called silica. One half of the earth's crust is silica, so it obviously plays an important role in farming.

The smallest soil grains are clay. When you rub soil between your fingers, the gritty, rough portions are sand and the smooth, sticky portions are clay. Silt is somewhere in the middle and feels velvety. A grain of clay can be a hundred times smaller than a grain of sand.

 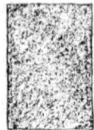 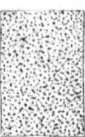

Crumbly Structure   Compact Clay Structure   Coarse Sand Structure

Sandy soils drain well because of their course texture. Although they warm up quickly in the spring and are loose, they lack essential minerals and thus fertility. Finely textured clay soils hold moisture, minerals and fertility, but are cold and compact when moist, and turn hard when dry. As the best soils are a combination, it is good to add clay to sandy soils and sand to clay soils.

Any substance formed by plants or animals is called organic matter, which differs from the mineral matter in that it comes from life. The life in the soil is in the decaying plant and animal portions of the soil. Humus refers to the black, waxy complex substance coating the soil grains that is made from the decaying organic matter.

It takes soil microbes, starting with bacteria and fungi, to turn organic matter into humus. Soils well supplied with humus are the easiest to farm. The farmer's concern for the soil quickly turns to care for certain soil microorganisms. Good farming practices promote the life in the soil.

Glaciers played a significant role in powdering rocks and leaving deposits of organic matter. This made for the great black soils of the midwest. The southeast did not get any recent glacial activity and has some of the oldest, most eroded soils in the world. Except for alluvial soils along flood plains where soil from upstream is deposited, most soils in the southeast must be improved before farmers can grow crops successfully.

Improving Soil—As observation is a key to learning, closely comparing a handful of rich garden soil with one from a worn out field can teach us a lot. The garden soil, with its additions of organic matter and minerals, will be dark and crumbly, while the worn out soil will be lighter in color and compact. The difference is that the former has humus.

Lichens and mosses are the first plants to grow on newly weathered rocks. As they grow, acids leaching out from their roots further decay the rocks, slowly creating and improving the soil. Higher forms of plants can then grow and the process of improving soil slowly continues.

Lichens obtaining a foothold on rock

Animals eventually enter the picture, eating plants and excreting. Their waste products speed up the soil improvement process. Acids in these wastes continue dissolving the minerals in the rocks which then become nutrients for further plant growth.

Worn out soil can be made into good garden soil, but only plants can do it, and they need the help of animals. Another difference is one we can't see, and that is the number of living organisms that can only be seen with a microscope. A spoon full of garden soil can have billions of these microbes, where the worn out soil has just a few million, or 1,000 times less.

Farmers rest their worn out fields by sowing them back into grass and clover. The immense network of the grass roots subdivides the cloddy soil into smaller crumbs, and the clover roots dive deeper down and bring air into the soil. You may have noticed that good soil has air pockets in it, while worn out soil does not.

The life in the soil, just like you, needs air to breathe. Air is made of the elements nitrogen, oxygen, and carbon dioxide. Both plants and animals cannot live without these important elements. As the sod grows, it opens up the soil so air can enter. Nitrogen is usually the limiting factor in plant growth and other chemical reactions, and oxygen is necessary for combining with the mineral elements so they can be of use. Plants incorporate the element carbon into the soil through photosynthesis, which is the process of converting the carbon dioxide in the air into the carbohydrates in the plant.

When the soil is open, water can enter instead of running off. Water is a combination of two elements, hydrogen and oxygen. Careful tillage opens the soil, but too much will damage the precious soil life. The water and air in the vicinity of a plant's leaves and roots are the source of the four free elements: nitrogen, oxygen, hydrogen, and carbon. Farmers do not have to add these when they routinely recover their cropland with a grass and clover sod.

Another great asset of grass and clover is that cows, sheep and goats love to eat them. Their stomachs contain different microbes which end up in the soil and help it get better. These animals are called ruminants, and are capable of improving soil even while living off the plants grown there.

Many other smaller animals help as well. Ants bring to the surface the finest sand. Earthworms make channels and take organic matter from the surface deeper into the soil, pulverizing it as they go. Birds, reptiles and mammals are all poking around the soil, moving minerals and organic matter around, and leaving valuable soil-building wastes behind.

The acids in plant roots and animal wastes continually help decay rocks and release mineral nutrients. The major mineral nutrients are the elements silica, calcium, potassium and phosphorus. Other minor but vital elements are sulfur, magnesium, iron, aluminum, sodium, copper, zinc, manganese, molybdenum, cobalt, chlorine and boron. These are known as the trace elements. Most soils contain enough of the major and trace elements, in an unavailable, insoluble form, to grow abundant crops for thousands of years, with the possible exception of phosphorous. If a soil is deficient in a major or minor element, farmers need to add it.

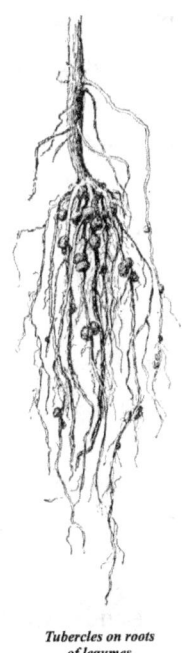

Tubercles on roots of legumes

By spreading substances with calcium, magnesium and potassium, such as lime and wood ashes, a chemical reaction occurs. These bases react with the acids to speed up the soil improvement process and release of nutrients. They are necessary for the growth of legumes, which is the family of plants clover belongs to.

Once the nutrients are released from their parent materials, the rocks, they can be leached out and washed away when it rains. We need microbes to incorporate loose nutrients into their bodies to prevent leaching, and now we see why the life in the soil is so important.

Farmers do many different things to make the land fertile and in good tilth. These things don't work as well by themselves. By growing grass and clover, raising livestock and spreading compost, wood ashes and rock dusts, farmers take great care of the life in the soil so they can grow healthy crops.

Compost—We've seen how nature slowly builds soil, how plants and animals help, and the necessity of life in the soil to release and then hold the nutrients. Putting all of this information together helps us understand why farmers make compost. This is where the microbes grow, reproduce, and form humus.

The grass and clover which a cow eats stays in her stomach (of which she has four) for over two weeks. It undergoes great changes there and comes out teeming with life and the possibility of fertilizing the soil. But life is fleeting and can go away quickly.

When you smell manure, you are smelling ammonia gas, a combination of nitrogen and hydrogen, which is leaving the manure and going back into the air. Farmers stop this waste by adding bedding, some carbon material like old hay or straw, to the manure. Now the nitrogen has somewhere it would rather be, because it loves carbon.

The third ingredient to the compost pile, besides the carbon material and manure, is soil. Soil has clay in it that aids in the production of the clay/humus complex we want our soils to become. Soil also has a different set of microbes, and we will soon see why "the more the merrier" is the theme for the microorganisms in the compost pile.

Sometimes farmers add a fourth ingredient, rock dusts. These are mineral-rich, ground up rocks from different regions of the United States. Granite, rock phosphate, greensand and other rock dusts add valuable minerals which the microbes digest and release in a more readily available form.

Compost is often made where the farm's livestock are fed. During spring, summer and fall, the livestock are moved frequently to different pastures to graze and fertilize the soil. The other pastures grow back with no animals on them. Hay is made when the grass grows quicker than they can eat it, usually in May and September.

Compost pile

Farmers harrow in the droppings and surface organic matter after the animals leave a paddock. This creates sheet composting that happens right in the field. During winter the hay is fed to the cows, and in those places we have the three ingredients for making compost; hay, manure and soil. The soil there has also been peed on, and the urine of livestock is every bit as valuable as the manure.

Many different species of fungi live in the forest, so adding leaves to the compost pile is beneficial. It's better if they are chopped up or decayed, as they will tend to pack in layers otherwise. After wood chips have sat for five or ten years, they will be decayed enough to benefit the compost pile. If they still resemble wood chips, they have too much carbon and any nitrogen will be used to further their decay, rather than for growing crops.

In the old days with a pitchfork, and now with a front-end loader, loose piles were made and moistened so the ingredients could ferment together. The compost pile will get warm and break down the materials so that in a year nothing original will be recognizable. Although some nitrogen, phosphorus and potassium will be directly available when the compost is spread, the real benefit is the untold number of microbes, ready to go to work improving the soil for the next crop.

How Plants Grow—A seed contains stored up nutrition. It will either be eaten by an animal or human, or it will sprout and grow or simply decay into the soil. When it sprouts, the stored up nutrients feed the seedling until it gets its first true leaves.

A bean seed swells with moisture when it is firmly contacted to the soil. Its neck curls up through the surface and two fat, oval leaves open up that each look like one half of a bean. These are called the "seed leaves" and are what's left of the bean still nourishing the seedling.

Soon the first true leaves appear, and the tiny root gets little hairs on it. Then photosynthesis takes over. The underside of the leaf has small gateways that open and close. When they are open, air goes in and excess moisture goes out. They close up when the plant needs to conserve moisture.

Sunlight is the power that drives photosynthesis. The leaf breathes in carbon dioxide and exhales oxygen. Plants and animals work in harmony because animals and humans breathe in oxygen and exhale carbon dioxide.

Bean seed on left, young plant on right, young plant with first two true leaves in center

Water is drawn up from the soil by the root to replenish the drying effect of the wind and the sun on the leaves. Plants have the ingredients, carbon, hydrogen and oxygen, they need to make carbohydrates such as sugar and starch. But they also need nitrogen. Although the air is about 78% nitrogen, plants can't use this directly. This is where the microbes play their important role.

As the roots of certain species of plants penetrate the soil, unique carbohydrates flow out from them. Specific species of dormant microbes that feed on these carbohydrates are awakened and begin to grow and propagate. Some of these are capable of extracting nitrogen from the air in the soil and making it available for the growth of plants. Other microbes make available the locked up minerals in the soil, such as potassium, phosphorous, calcium and the trace elements.

Because these bacterial and fungal microorganisms are dependent on plant growth for their food, they have a vested interest in the health and growth of the plant. Fungi have long roots stretching throughout the soil and connecting the plant roots to the rest of the garden soil. There is no hard and fast line separating the end of the root and the soil.

Microbes breed fast when food and water are available, so they are constantly giving the growing plant whatever it needs. Plant growth is governed by sunlight, the atmosphere, moisture, and the soil with its microbial life. These are the four classic elements of fire, air, water and earth.

The Elements Farmers Work With

Sunlight— Sunlight warms up the earth in spring, and plants then grow. Seeds need different soil temperatures to sprout. For example, lettuce seed will sprout when the soil is 50 °, but not when the temperature of the soil is over 75 °. Lettuce likes cool weather.

Corn, a warm weather lover, won't sprout when the soil is below 55 °, but prefers 65 ° or higher. Although spring crops can be planted "as soon as the ground can be worked", it is often better to wait until the soil warms up. The faster the seeds germinate and grow, the easier it will be to control weeds.

On a hilly farm there is a noticeable difference in the growth of certain crops when planted on top of the hill or down in the valley.

In the low places the sun rises later and sets earlier. The sun-loving, carbohydrate-producing crops, like tomatoes, peppers, squash, melons and sweet potatoes won't produce as well in the valley as they will up on the hill. Greens like lettuce prefer the shadier valleys.

Another factor is the proximity of trees and shrubs. Particularly in urban or suburban environments, large trees are needed for shade. Gardens won't thrive well when shaded for part of the day. Ambient light can be increased by thinning out trees, shrubs and branches along the edges of gardens. You may not notice the encroaching branches that yearly extend and thicken along the woods line. Thinning not only adds more light, it helps to create better air flow.

It is interesting to note how the angle of the sun shifts throughout the year. At the spring and fall equinoxes, approximately the 21st of March and September, the angle of the sun from the horizon at noon will be 90 degrees minus the latitude where you are located. At the equator the sun is always directly over head at noon. Every place on earth gets 12 hours of sun on the equinoxes.

As spring progresses in the northern hemisphere, the sun gets higher in the sky until the summer solstice, approximately June 21st. This is the longest day of the year. Then the sun gets lower in the sky until about December 21st, the winter solstice, which is the shortest day of the year. The moon mirrors the sun's annual movement by riding highest in winter and lowest in summer, in monthly cycles.

Many plants are sensitive to this cycle of light and won't begin flowering until the days start getting shorter. Humans are emancipated from those cycles but animals are more sensitive to them. Plants are totally dependent on light and warmth and consequently only thrive in their proper season.

The Atmosphere— Most of the elements that make up a plant come from air and water. They are nitrogen, oxygen, carbon and hydrogen. These four elements make up 95% of a plant. The air pores in the soil form a very important part of the soil. This is why farmers concern themselves with tilth and humus.

An ideal soil would have twenty-five percent of its volume be air. The small crumbs or grains of soil would be separated from each other by air pockets. Humus plays the role of keeping the soil fluffy and in good tilth.

Heavy rains, tractors and footsteps all compress the soil and compact it. After every rain that presses down the soil surface, the farmer is anxious to fluff it up again so that plant roots have the air they need. As soon as the ground dries up enough so that a handful of soil squeezed together shatters when it is dropped, the farmer cultivates and the gardener hoes.

Cultivating tractors have shoes behind the rear tires that immediately fluff up the soil that the tire just compressed. They are called middle busters because they break up the ground that is in the middle of two rows. Gardeners do not step on ground they have just hoed, but work from the side and hoe where they have stepped.

The air pockets in the soil fill up with water when it rains. This water needs somewhere to go. The first thing to consider in growing plants is drainage. Land that puddles and stays wet after a rainstorm needs to be drained.

Some land requires tiles underground to remove excess water and allow for ventilation. Air flows back up tiles when they are not full of water. Tiles are usually trenched in around three feet deep and slope gradually to a drainage ditch.

Farmers use a subsoiler in dry weather to make two foot deep trenches to help with drainage and underground air flow. Gardeners make double dug beds by removing the topsoil in a trench, and then breaking up the subsoil a foot deep with a pick or digging fork. Often they will add sand, limestone or gravel to the clay subsoil to lighten it up before pushing the next layer of topsoil in the trench. The first topsoil removed gets put back in the last trench of the bed.

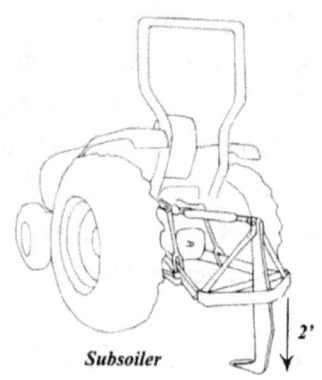
Subsoiler

Cover crops let nature put air into the soil and keep it open. The massive root systems of grasses and grains create great soil structure, and the deep taproots of legumes open up the subsoil and bring in the atmosphere, with most of the elements necessary for plant growth.

Moisture— Water in the soil comes in three forms: free water, film water, and capillary water. Free water fills up the soil after a rain and then sinks down to the water table, which varies in depth throughout the year. Plants cannot use free water because it has filled up all of the air pockets, so if the water table is shallow the land must be drained.

Film water sticks to the surface of the soil grains by adhesion, like moisture on a stone dipped in water. A good soil can hold one half of its weight in film water, and this is what the plants use. As it is used it is replaced by the third type of soil water, capillary moisture.

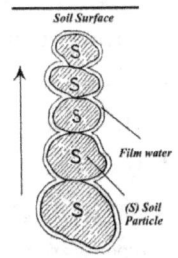

As plants use up moisture near the surface, subsoil moisture rises upwards along film water by capillary action

Capillary action is the movement of water from moist soil to dry soil, regardless of gravity. If you dip paper into a cup of water, the rise of moisture up the paper is called capillary action. Crops rely on it, and the farmer's part in preparing the ground and cultivating the soil consists chiefly in controlling the movement of capillary moisture.

Farmers do this by plowing to loosen the soil and then allowing it to settle. As the crop grows the surface is stirred. The large air pores thus formed on the surface prevent the moisture that's a little deeper from rising up and evaporating.

Gardeners often use a mulch to prevent evaporation, where the farmer uses the soil itself as a "soil mulch". As plants use up the moisture near their roots, capillary moisture moves free water from deep below through the film water surrounding the soil particles to the drier places around the roots.

Now we can see why tilth and organic matter are so important. Rains need to soak in deeply and then be prevented from evaporating. Fall plowing, contour tillage, terracing and loosening up compact soil and subsoils are methods to permit more rain to soak into the ground, along with the practices of using compost and cover crops to create humus and good tilth.

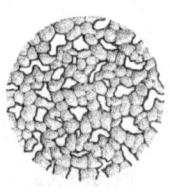

Air spaces, white
Soil particles, shaded
Film water, black lines

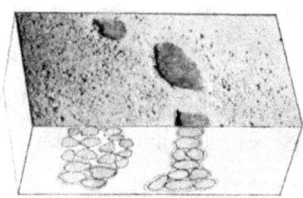

The footprints are kept moist because the soil moisture is permitted to move to the surface, which can't happen when the soil is cultivated

When you step on cultivated soil you leave a footprint indentation of an inch or two. Notice how it stays moist while the soil around it is dry. Picture a bunch of straws going down from the footprint into the free water at the water table underground. Moisture constantly moves up these straws by capillary action into the dry air and wind above the soil.

Surface tillage or a mulch plugs up these straws, conserving the moisture in the soil. Farmers want the water to evaporate only through the transpiration of plant leaves. On hot windy days a soil can lose an inch and a half worth of rain in a week. By simply mulching or lightly tilling the soil surface this moisture is retained and then later used by the crops.

Soil Microbes— Germs and molds have a bad reputation because some of them can make you sick. But let's not "throw the baby out with the bathwater". There are over 100,000 different species of bacteria, and 1000's of different species of fungi, and they play an essential role in life and death.

You have DNA in your body that is unique to you. All living things do. But there are other living things inside you that have DNA that is not yours. Microbes outnumber the cells containing your DNA by 10 times. Ninety percent of the DNA in your body is not yours.

This is a good thing because these germs help us digest food, ward off diseases and keep us healthy. After a round of antibiotics (which means against life) we are advised to eat microbial rich foods with probiotics (for life) such as yogurt. The same thing happens in soils when we use compost to reintroduce various microbes there.

By layering mineral-rich soils with organic materials in the compost pile, bacteria and fungi propagate and proliferate. When this is liberally incorporated into the soil it coats the soil grains with humus. We could say that it makes the soil intelligent because the microbes receive signals from the plant about what the plant needs.

The good germs in the soil need an abundance of vegetable matter to feed on and the addition of lime to keep the soil from becoming sour. They thrive in warm, moist, well-drained soils which are kept loose so that air can circulate in it. The bad germs grow in compacted or wet soils, or in soils deficient in organic matter.

The fungal roots are coated with silica, and they are like tubes winding their way throughout the soil. To get an idea of this, picture the aspen- covered, Colorado mountains where the trees are all interconnected underground. A dye put into one tree will show up hundreds of feet away in a day, traveling through the underground fungal connection.

Now we can see the important role silica plays. Like a crystal, it doesn't dissolve. So the nutrients flow through the tubes of fungal roots connected to the plants we want to grow. Although bacteria and fungi are the primary feeders in the soil, releasing locked up minerals, there are plenty of other living beings in the soil that feed on them.

The protozoa that eat them are in turn eaten by other life, and it continues on up the food chain to earthworms, bugs, reptiles, birds and mammals. The minerals needed by the plant become part of the bodies and waste products of all of these beings. Earthworms also take surface litter into their channels, mix it with soil, and leave minerals in a non- soluble but easily available form in their castings.

These minerals eventually wind up in the plant through what we can call the intelligence of nature.

Nature. — When soils have billions of microbes in every spoonful, they are in good shape to grow healthy crops. If soils have 1,000 times less, crops will suffer.

Nature knows best. The soils lacking microbes are the soils needing humus. Humus comes from decaying organic matter. So these soils need organic matter that was once alive to be dead, so that it can decay and form humus.

Plants are always sending signals. A plant growing in a soil lacking humus and these great quantities of microbes sends out a certain signal. These plants, if dead, could help build up the soil humus. Nature responds to this signal and sends in her house cleaners.

In the household of nature, the cleaners are the leaf-eating bugs and the plant diseases. They are the beginning of the humus-forming process. Farmers avoid these problems by beginning the humus-forming process themselves, in the compost pile. Plants grown in a humus-rich soil do not send these signals out that invite these problems.

Still, there is the issue of getting nutrients to the growing plant. Let's say a plant needs potassium. Potassium strengthens the stalks of plants, and our plant is starting to lean over from a lack of potassium. A signal is sent saying, "I'm going to fall over if I don't get some potassium".

This worries the soil microbes living off of this plant's root exudates, which are the carbohydrates flowing out of the roots. These microbes are going to die if the plant falls over. So a carbohydrate is sent out through the silica-coated, fungal root tube and a hydrogen ion is traded for a potassium ion that is on a particle of wood ash that the farmer spread last winter.

As the potassium ion is headed back up the tube towards the plant a disaster can happen. The farmer, trying to conserve moisture, aerates the soil by cultivating the crop and slices through the fungal tube carrying the potassium that the plant needs. The plant leans more.

Nature comes to the rescue. An earthworm eats the fungus and heads towards the leaning plant. But when it comes to the surface a bird eats it and flies up into a tree. Then the bird excretes and the potassium ion gets picked up on a mouse's foot from under the tree and the mouse heads towards the plant.

But then a snake eats the mouse. Soon the snake slithers over to the leaning plant and sheds its skin. The potassium ion, still involved in some microbial relationship, finds its way from the snake skin and into the plant. The plant straightens up and this is a simplified example of how nature works. Nature is extremely complicated and everything is interconnected.

Plants and Their Parts

Leaves—Plants grow from leaves. The leaf gathers light from the sun and carbon dioxide from the air, and pulls up water from the soil to make carbohydrates during photosynthesis. Through the sap flow, the carbohydrates give the plant the energy to grow downward making roots and upward making more leaves and eventually a flower.

When a leaf starts to wilt from the sunlight, the tiny air passages close and moisture flows upwards from the water in the soil. Sap pressure and capillary action are strong enough to pump water up to the top of the tallest trees.

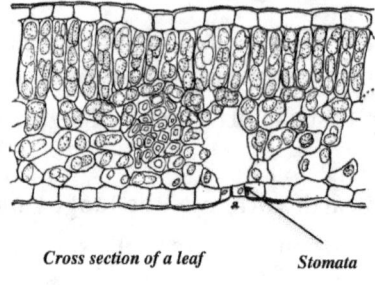
Cross section of a leaf   *Stomata*

The door for letting air into the leaf is on the underside and is called the stomata. Leaves like a gentle breeze to bring them fresh air. Farmers know how big their plants will get and space plants accordingly or thin them out later for good airflow.

Leaf cells contain chlorophyll, which absorbs energy from sunlight for photosynthesis. Leaves look green because of chlorophyll. Leaf cells are similar to our blood cells, but instead of iron in them they contain magnesium. If magnesium is deficient, farmers use a high magnesium lime called dolomite. A gardener might use Epsom salts, which is magnesium sulfate.

Some of the carbohydrates that flow down into the roots are released into the surrounding soil. Because it exits the root it is called a root exudate, and its composition is unique to that plant's family. These root exudates are the food source for the specific beneficial microbes that attach themselves to that root and help feed the plant.

The first true leaf follows the seed leaves, or cotyledons. It is rather simple at first but the next leaves get more complicated until there is the form typical for that plant species. If there is an excess of soluble nitrogen, the plant will keep growing more leaves.

Remember that nitrogen is the limiting factor in plant growth and most other chemical reactions. So as the season progresses nitrogen will be used up. The next leaves will be smaller and thinner, eventually forming a ball at the top of a stalk.

This ball is the bud of a flower. It is wrapped in simple looking leaves called sepals. They might start to show a little color. But the real color comes from the petals which burst forth when the bud opens up.

On the other hand, trees often flower before the leaves come out in the spring. That is because they made their flower buds the year before. When fall arrives with less light and cooler temperature, the green chlorophyll production stops and the other colors in the leaf become visible.

Flowers— Except for mushrooms and simple plants, all plants have a flower, from the tiniest weed to the biggest tree. The flower petals expand from the sepal covered bud and attract attention. Flowers often have an aroma which also attracts attention, and the most attentive ones are insects.

Inside of flowers are the organs of reproduction. Along the axis of the stem is the pistil which has the ovary at its base. This is the female part, and up on top of the pistil is the stigma. The stigma is sticky and the pistil is a long tube going down into the ovary.

Surrounding the pistil and stigma are a group of thin slender columns called styles, and on top of them are the pollen covered anthers. These are the male parts of the flower, and the dust-like pollen is the finest, tiniest part of the plant.

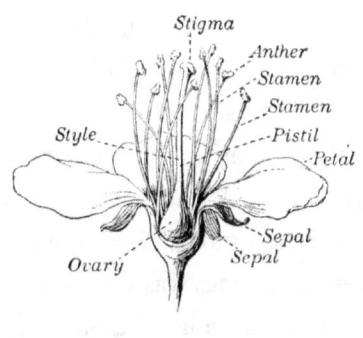

Parts of a flower

Pollen can blow in the wind, and many plants are pollinated by wind-carried pollen dust. But most crop plants require pollinating insects to come visit them. A farmer can take good care to ensure these insects are present by growing crops for them or raising bees.

To prevent inbreeding, most plants need the pollen from another plant in their species. Insects visit flowers to gather nectar and then pollen gets on their bodies. When the fly off they visit another flower of the same species and some of the pollen falls off them and onto the sticky stigma.

Now it gets drawn down the pistil tube into the ovary and fertilization takes place. When this occurs the flower fades and the plant produces some kind of fruit with seeds. The energy of the plant is then exhausted.

Farmers and gardeners know this and so they must keep certain crops harvested before they make a mature seed. These are the crops we eat at an immature stage, like cucumbers or summer squash. Once these plants make large, mature fruits they lose the energy to keep flowering and fruiting.

The flowering process relies on phosphorous, another element that can be limiting in agricultural production. Phosphates are mined and spread on the land as a ground rock dust. Phosphates can be made more available by adding hydrochloric or phosphoric acid, but then they tend to leach away into the ground water and eventually become a pollutant in rivers.

There are soil microorganisms that can mobilize the insoluble phosphate in the ground rock and make it available to the plant. These organisms are called into action when the plant needs the phosphate for flowering. Besides waste and pollution, water soluble phosphates destroy these particular microbes and then the soil is no longer able to do this process

Many crops are grown just for their flowers. From the pretty annuals like zinnia and sunflowers, to bushes such as roses and lilacs, a farm usually has flower beds. Bulbs are planted in the fall for early spring blooms. Besides attracting beneficial insects, colorful flowers add beauty to any place.

Seeds— Seeds are formed after flowering. The seed must contain all of the nutrients the new plant will need to sprout and grow up out of the soil to make the seed leaves. Only after the seed leaves open will photosynthesis take over the job of growing the plant.

The seed must also have the ability to resist destructive forces until it is time to sprout. So seeds are extremely hard, besides being the most mineral rich and nutritious part of the plant. Most animals and humans rely on seeds in their diet.

Seeds are harvested when mature and dry. They will sprout when moist. Special storage areas are needed to keep seeds thoroughly dry and viable.

Some seeds won't sprout unless they have been frozen. This is the way nature ensures they don't sprout in the fall and then die in the winter. Stratification is the process of freezing seeds to get them ready to sprout.

Not all seeds are going to sprout and grow. A mustard plant can make a lot of seeds. If all of them sprouted and grew, and this happened for five generations, there would be enough seed to plant them one foot apart over a sphere whose circumference is the earth's orbit.

There are a lot of tiny seeds already in the soil. Charles Darwin, a scientist in the 19th century, sprouted over 500 plants from a spoonful of soil. Farmers know how quickly bare soil fills up with weeds, and are always eager to keep the crops they've planted the only ones in the fields. Once a weed gets established it sends signals into the soil to sprout other weeds in that family.

Seeds are moved around by a wide variety of ways. Many are inside tasty fruits which are eaten by animals and excreted elsewhere. Some are lightweight and get blown about by the wind. Others have sticky coatings and are moved around on animal fur. Squirrels are good planters of tree seeds.

Farmers have saved seeds since the beginning of farming, often selecting from the very best plants. The history of corn is a fascinating study of how a small grass seed was made into a big ear by human selection. The crops farmers grow now were improved over many years from their wild ancestors.

The genetics in a seed comes from both parent plants, the one the seed is collected from and the one that supplied the pollen. Special crosses can be made by humans taking pollen from one plant, putting it on the female part of another plant, and excluding all other pollen. A hybrid is a specific cross between two different varieties of a species which results in a plant with known characteristics. But seed taken from the hybrid will not necessarily become like the parent because of the genetic diversity of all of its ancestors.

Seeds that have been saved for a long time are called heirloom seeds. They became adapted to the area where they've been grown. Seeds that have had their genetics artificially modified with the DNA of a completely different species are known as genetically modified seeds.

Plant Propagation— Plants propagate themselves by either seeds or buds. Fruits do not often come true from seed because the seeds are formed by cross pollination and carry the genetic diversity of many ancestors. So many fruits are propagated from buds instead.

Seeds are used to grow root stocks for the new plants. Then a bud, or a twig with buds, is put on the stock. Some plants can be propagated by simply taking cuttings, or by layering.

An apple seed will likely grow a crabapple, but out of a large number of crabapples a tree with good fruit may be found. Its characteristics might include tastiness, large size, or disease-resistance. A farmer can take a twig of an apple tree and make identical trees, by making a whip graft.

One year old shoots supply the twigs, or scions, in the fall. During winter the trees are grafted with a very sharp knife. A smooth cut is made at an angle a few inches above the roots of the stock, and then a vertical slit is made down the center.

A similar cut and slit are made on the scion, which should contain 2 or 3 buds. The two are spliced together with the vertical slits interlocking them, being careful to align the cambium layers together. The light green cambium, directly under the bark, is where the cells divide and is the actively growing part of the tree. The cambiums of both rootstock and scion must line up with perfect contact, and then the whole graft is wrapped with masking tape, or tied with yarn dipped in wax. These are stored until spring in a cellar and then set out in the garden for a year before planting in the orchard.

Whip Graft

Budding is done by slipping the bud from the desired tree into a rootstock during the growing season. The leaf stem is left on to use as a handle for slipping it under a T cut made on the rootstock. This is done in the field, and if the bud sprouts, then the upper part of the tree is pruned off.

Pome fruits are usually grafted and stone fruits are budded. Raspberries can be tip layered by burying the tip of the cane in the soil, where it will form roots. A root of a raspberry will have buds on it, so cuttings of the roots can be dug and set out elsewhere to make new plants.

Layering is a way to propagate vines, like grapes. A growing branch is buried and a bud on it will make roots. A mound of earth thrown up around some shrubs, like gooseberries, will make the lower branches form roots. These can be cut off and planted out the next spring.

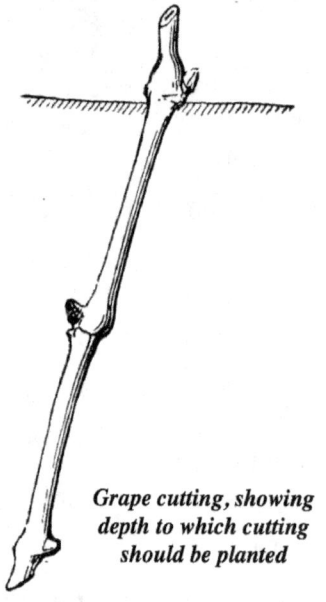

Grape cutting, showing depth to which cutting should be planted

Cuttings of the ripe shoots of some plants, like grapes and currants, are made about 6 inches long in late winter and buried in damp sawdust in the cellar. In spring they are set in a deep furrow with only one bud above the soil. Some plants, like rhubarb and many herbs, are simply dug up and divided into several plants.

Steps in Budding

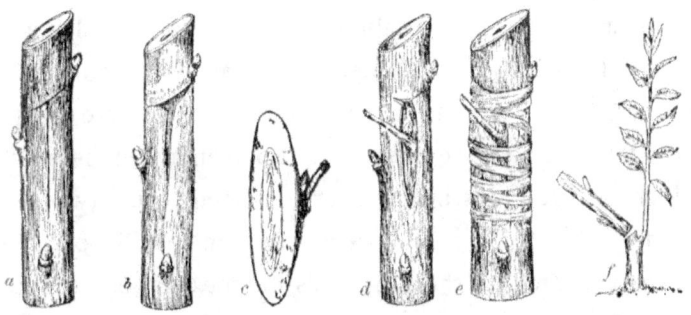

a. *Cuts in Stock*
b. *Bark Slipped Away*
c. *Bud*

d. *Bud Inserted*
e. *Bark Wrapped*
f. *Bud Making New Tree*

The Care of the Soil

Creating Humus-Rich Soil—There are no recipes for farming. Soils are different, crops are different and farmers are different. All we can do is try to understand nature and mimic her the best we can.

Market gardening requires the most compost of all types of farming. Forty to fifty tons of well-rotted compost per acre per year is not uncommon for garden crops. If the soil is naturally fertile it may not need as much.

Compost and other organic materials are acidic, so alkali substances, such as lime and wood ashes, are also required. The soil PH, tested with litmus paper, should be around 6.8 for most crops.

Lime varies in its composition. If it has a high magnesium percentage it is called dolomite. This is used when soils lack magnesium. But if the soil is sticky it is high in magnesium and then a high calcium, low magnesium lime is better. It is spread at the rate of one ton to the acre every two years, or 50 lbs. per 1000 square feet.

Wood ashes are also alkaline, containing both calcium and potassium along with trace elements the tree has brought up from deeper soil levels. It is spread at about 400 lbs. to the acre every two years or 10 lbs. per 1000 square feet. The percentage of these cations (positively charged ions in the cation exchange capacity of a soil) would ideally be 65% calcium, 12% magnesium and 5% potassium.

The best way to turn these minerals and the compost into humus is by growing a wide variety of cover crops. In the summer, buckwheat, cow peas and beans are the best. In the winter, farmers plant a mixture of cereal rye and purple hairy vetch or winter wheat and Austrian peas. It is a good idea to add daikon radish or turnips to the mix. Grass and clover are used when the field won't be cropped for a few years.

Each of these crops have specific soil microbes that live on their root exudates, and play specific roles in unlocking certain nutrients. The grains and grasses work with silica. Buckwheat, which despite its name is not a true grain, helps free up calcium. Beans, peas, vetch and clovers bring in carbon, oxygen and nitrogen. The brassica family, which includes daikons and turnips, helps with sulfur and phosphorous. Growing cover crops in a field that has both the compost and cations incorporated into it prepares it well for market gardening.

Soils can be tested for their total and available nutrients. Ground rock dusts are good sources for phosphorus, potassium and the trace elements, as are kelp and other sea products. A continuous supply of the organic matter refuse from crop production, along with compost from the farm's animal manure, create a humus-rich garden. Other types of farming can get by on less fertility, but all crops thrive with high fertility and good tilth. The farmer's concern is for creating humus-rich soil, not the crop. Then nature will take good care of the crop.

Tillage—Tillage is what the farmer calls a "necessary evil". As we saw in our example, when the farmer cultivates the garden, fungal roots get chopped up and prevent the flow of nutrients. Any time we till we destroy the delicate interplay of the precious soil microbes. If we could simply mulch everything, as in a small garden plot, this would alleviate most of these issues, especially if we've double-dug the beds. Good mulches are rotted hay or chopped up, rotted leaves. Nitrogen will unite with the carbon in unrotted organic materials, temporarily robbing it from its job of growing the crop.

On larger gardens or fields tillage is required. Here are a few quotes about tillage from old time farmers. "Work the soil as little as possible", "Plow like you are turning over a sleeping baby", and, "Only till when the soil breaks apart easily". These sayings indicate the need for minimum, gentle and slow tillage when the soil has the proper moisture content.

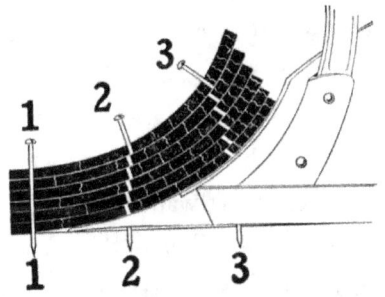
Shifting soil layers by the moldboard plow

The moldboard plow flips the soil up. The result is similar to the action of bending a paperback book where all of the pages shift on one another, doing an excellent job of loosening and

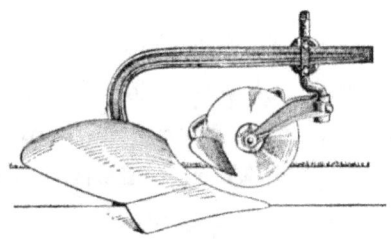
Moldboard plow

incorporating air. Farmers are quite interested in plowing and intensely observe the soil as it breaks apart. If the soil appears waxy or shiny and sticks together, it is too wet to plow and the land can be harmed. It can also be too dry and powdery to plow.

Fall is the best time to work with the moldboard plow for many reasons. Farming chores are not as pressing as in spring. The soil is often drier in the fall. The winter freezing and thawing of the upturned soil has a good pulverizing effect. Some crop pests are exposed and destroyed in the winter and surface litter has time to become incorporated.

When a wet, clay soil is worked, as in the case of plowing too early in the spring, the soil's crumbly structure is destroyed and the particles run together, forming a solid mass. When it dries up this mass gets hard and cloddy, and the soil has lost its structure, or tilth.

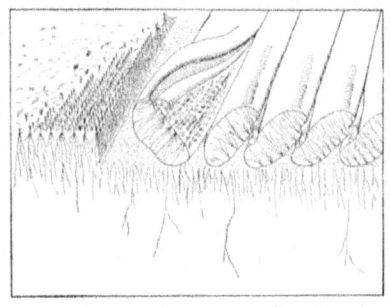

Sod turned 3/4 of the way over

The moldboard plow is usually set so it flips the sod about 3/4 of the way over. The depth is to the level where the subsoil begins, about 6 to 8 inches. Farmers don't want to bring the infertile subsoil up to the surface where it will interfere with crop production. The toplink, one of the implement arms, and the plow itself are all adjustable. The plow shoe should be level at the depth the plow is set for. The coulter slices the sod above the plow.

Each furrow completely moves the sod on top, without leaving a grass strip. A problem with the moldboard plow is that by flipping the soil over, the intricate layers of biological activity that are so beneficial become inverted. They re-establish themselves, but this takes energy and time. The moldboard plow is only used on sod or on really compacted land. Otherwise farmers use a chisel plow.

The chisel plow does not invert the soil, and whenever possible, is the first choice. Springs located horizontally alternately tense and release so that the shanks vibrate forward and backward with a chipping action. This breaks up the soil under the surface and cracks it laterally. You can tell how much organic matter is in the soil by the ease of plowing and the way it breaks apart. The fine roots of grasses and other organic material prevent the individual grains of soil from running together.

In all tillage, moving slowly and gently reduces the damaging effects. By waiting a few days between the primary chisel plowing and subsequent tillage, the soil microorganisms reorganize themselves and send signals about the change of plant species that is happening. Plants die easily when their specific microbes are no longer helping.

A cover crop will have a thriving microbe population helping it to grow and they are at their peak right before flowering. When this is mown and chisel plowed, root exudates stop flowing and many microbes specific to this crop die. By waiting a few days, the microbe population shifts, and the crop decays quickly. Farmers should not be in a hurry and shouldn't go over the ground several passes at once. It is much better to let time and microbes help do the work of tillage.

The tool farmers use on the final pass is the spike-toothed harrow, which has rows of 6" long teeth that are offset. It smoothes the clumps and creates a fine seed bed like a rake does in a garden. If the soil has too many of the bigger clumps, a disc harrow is used first to chop them up. On clay soils a disc can make the clumps worse if they are at all damp. Any tillage should only be done when the soil breaks apart easily.

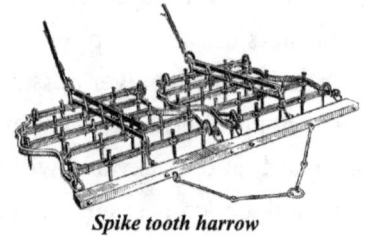
Spike tooth harrow

Farmers are always concerned about soil moisture. Plowing and harrowing conserve soil moisture unless it's done when the soil is too wet. In that case they create clumps and do not leave the land in a good condition to prevent evaporation.

Another use for the harrow is to spread out manure after a paddock has been intensively grazed. Cows don't like the nitrates that accumulate around their waste and won't eat the dark green circle of grass around their cow pies. But this valuable nitrogen source becomes beneficial for the regrowth of the pasture when it is spread out and mixed with a little soil.

A third use of the harrow is to create a stale seed bed, or to drag directly over a newly sown crop. Bare ground will not stay bare for long. Nature abhors a vacuum and weed seeds will quickly sprout. By dragging the harrow over the seed bed instead of planting, and/or right after planting, tiny sprouting weeds are easily destroyed before they have emerged. The seeds that the farmer planted are deeper in the furrow and are not disturbed.

Some crops, such as potatoes or corn, can be harrowed even after they have sprouted and not be damaged. This greatly reduces weeds sprouting in the row that subsequent cultivation can't remove. All of this harrowing continues to conserve moisture by preventing evaporation.

As the crop grows, the soil between the rows must be cultivated to conserve moisture, aerate, and prevent weeds from establishing themselves. Rototillers chop up the soil too much and this destroys the structure. Shanks with narrow shoes that vibrate are quite sufficient and leave the soil in better tilth.

At first the tiny seedlings are in danger of getting covered up by the moving soil, so the farmer will have to go real slow. As the plants get taller, a little speed throws soil around the base of the plant, which smothers weeds. Some crops, like potatoes and corn, thrive with soil hilled up around them. Others, like onions and beets, would rather have their shoulders exposed.

When the plants are too big to get over, the crop is said to be "laid by". Weeds may need to be pulled or cut. Inter-sowing a different crop can prevent weed growth, but it should not interfere with the present crop.

As soon as the crop is harvested, the farmer sows a new crop or a crop grown just to improve the soil, called a cover crop. The land is never left to grow up in weeds. Market gardeners often plant two or three crops and a cover crop in the same year, always keeping their rows full of plants they choose.

Grass— The two major groups of plants are classed according to their first leaves. If they sprout up with one leaf they belong to the monocotyledons, and if they start with two leaves, they are dicotyledons. Further divisions of the families of plants are done by their flowers.

Grasses are monocotyledons. The leaves of this group split lengthways from the tip to the base. They have fibrous, shallow roots that are excellent at dividing up the soil into smaller crumbs. Grasses are great at creating good tilth and soil structure.

They also are the preferred food of the ruminant animals who make the soil more fertile. Any effort at soil improvement will include grasses. When they are cut or eaten, some of their roots die and decay, forming humus underneath. When they regrow, the roots grow a little deeper into the soil and the process of soil formation continues.

An indication of the value of grass is found in the old land lease laws of England. Out of every four years of use, the land had to be in grass for two of those years. They also had laws preventing the sale of hay, which would deplete the soil. The hay had to be fed to livestock so that the fertility stayed on the farm.

A diversity of grasses ensures animal fodder for each season. Of the many kinds of grass, some are better at soil improvement and others are more nutritious for livestock food. Some make edible seeds, like corn, wheat, rye, barley and oats.

Grasses are benefitted by growing clovers or other legumes with them. Legumes have nodules on their roots formed by a bacteria that can access the nitrogen from the air in the soil. Grasses can't do this, but can use the nitrogen formed by the clover roots. Clover's deep root does not compete with the shallow roots of grass. As companion plants they are symbiotic, which means together they create more growth than if they grow separately.

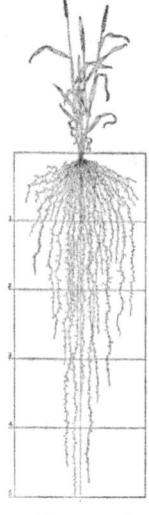

Roots of wheat plant

Clover and grass seeds are already in most soils. If you quit tilling and mow, you'll probably get a good stand. Remember that clovers need a high pH, so if there is no clover in a field, lime is required. Phosphorous also needs to be available for clover to thrive.

Where land is sloped more than 7 or 8%, it is too steep to grow row crops because of erosion. This land should stay in sod and be hay or pasture. A

Terracing

good method for improving sloping land is by terracing and keyline plowing. Developed in the dry country of Australia, keyline plowing is a way to keep the rain that falls on the farm from leaving. A transit is used to find the level contour, and small trenches are made along these lines, maybe falling slightly away from the valleys. Rain that would otherwise go down the valley is redirected towards the ridge. Several days after a rain these furrows are still moist. The soil stores this moisture in the subsoil and it becomes available later when the weather turns dry.

By taking every other shank off of a chisel plow, a handy tool is made to create trenches two feet apart. It is pulled at a depth where it goes into the subsoil an inch or two. If it is following heavy grazing, a harrow behind it will scatter the manure and level off the small ridges that are formed.

As the topsoil deepens, the trenches can be made a little deeper. This method works well in conjunction with a rotational grazing system that allows the plants to recover well before they are grazed again. Thus, the roots are able to get into the subsoil and keep it open. Keyline plowing and subsoiling are done when the soil is dry, otherwise the damp soil would smear and form clods.

Crop Rotation. — Different species of plants give and take different nutrients to and from the soil. Growing the same crop year after year depletes the soil of certain plant foods. Traditional farming methods stress the importance of rotating the crops on the different fields of the farm.

During the middle ages, farmers didn't choose what to grow. The priests or noblemen followed strict crop rotations that had been worked out for centuries. These rotations of different crops with ruminant pastures ensured the land would remain fertilized and productive. Their minimal equipment couldn't destroy soil structure much.

In our age, farmers have much more freedom, and freedom requires responsibility. When a farmer first plows up the sod, it is like making a contract with nature. The farmer agrees to give back compost, minerals and cover crops, and be gentle with tillage, in return for taking crops off of the field.

Specific crop rotations are different for each location and type of agriculture. One example of a four-year rotation would be corn, beans, wheat with clover inter-seeded, and then clover and grass for two years. Notice how the farmer uses both the grass family and the legume family either following each other or together.

In market gardening, farmers alternate different species every year. Problems would arise if the farmer grew only potatoes every year. By growing potatoes, and then squash, then beans and then sweet corn, each species takes and gives back different nutrients. Cover crops are grown every winter unless a field needs to be left roughly plowed to get an early spring crop in.

This requires dividing the farm, or garden, into different fields or plots. They are called rotations because the crops rotate around the farm, not returning to the original field until the fifth year. A farmer could also have a rotation of three years, or five or more years.

Animals are rotated frequently to different pastures. The more animals on smaller plots for lesser time allows for more time before they return to the original field. This gives the pasture more time to recover, regrow and improve the soil.

Farms are healthiest with the more diversity they have. Woods and wetlands are very important sources of nutrients that are exchanged with fields by reptiles, birds and mammals traveling back and forth. Most life occurs at the boundaries between woods and fields, or the wetlands and fields. Ancient agriculture rotated woods with fields, but the method used, burning, destroys too much carbon to be recommended.

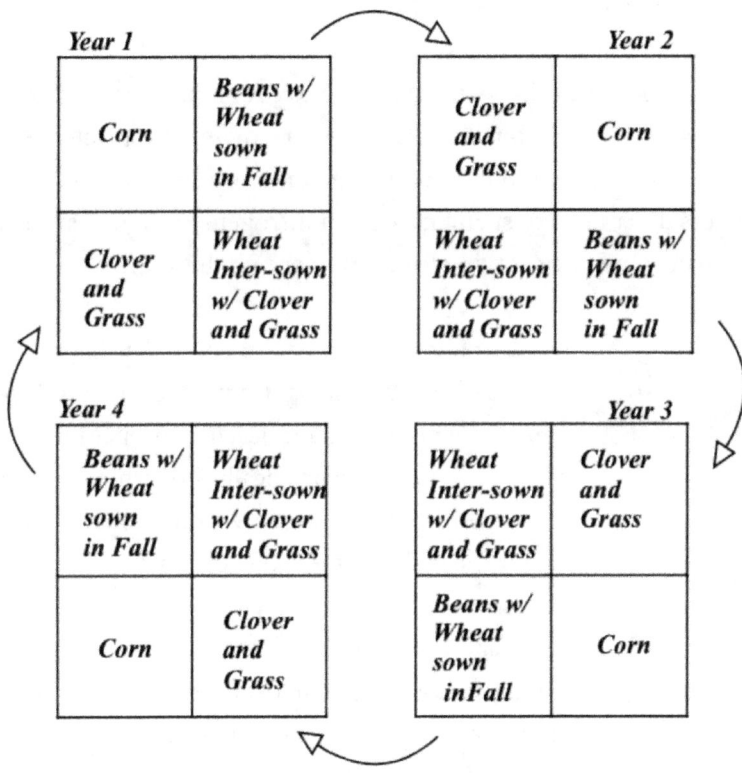

Showing how crops are rotated.

Year 5 same as year 1

Legumes— Legumes play an important role in the farm's crop rotations because of their ability to enhance the nitrogen in the soil. They do this through a symbiotic relationship with certain soil bacteria. You can see these nitrogen formations as nodules on the roots of legume plants.

Farmers can inoculate legume seeds with these bacteria by stripping off the root nodules of the same species and stirring it into the seed before planting. Soil from a field that successfully grew that species will contain the correct bacteria and can be added to the furrows about to be planted. Legume inoculants can also be purchased.

Each species of the legume family requires a different species of bacteria. These tiny fertilizer factories are the greatest help to the farmer. With 70 million lbs. of nitrogen in the air above every acre, farmers don't need to spend money on nitrogen. Each year a part of the farm should be growing legumes to enrich the soil.

Nodules on the roots of a young plant of garden pea

Legumes have deep tap roots that bring oxygen, carbon and nitrogen into the soil and bring up nutrients that have leached deeper into the subsoil. They also physically break up the hard pan below where the plow cannot reach. Many legumes make good food for humans and animals.

Legumes need a soil well supplied with calcium, potassium and phosphorous. Lime or wood ashes are used to sweeten the soil so that legumes can thrive. The soil must also be loose so there is air available for the bacteria to work with.

Clovers are the farmer's best friends. Red clover is good for hay fields and pastures. White clover is shorter and both are often mixed with grass seed when sowing the field. Crimson clover is an excellent cover crop to sow after the summer garden is harvested, but it is not a good stock feed. It is a biennial so it doesn't want to regrow after mowing the next spring. Sweet clover grows tall and has deep tap roots.

The vetch family has many good cover crop options. Purple hairy vetch is often mixed with rye or wheat, which it will climb. It must be mown in the spring before it goes to seed or it may become a weed problem. Garden members of the vetch family include fava beans and chick peas, and a perennial ground cover called crown vetch.

In the garden the legumes are the peas and beans. Austrian peas are a climbing legume often sown with rye or wheat. Shelling peas and snow peas are also widely grown. Cowpeas are a great summer crop for building soil organic matter and adding nitrogen. These include black-eye peas, crowder peas, whippoorwills and pinkeye purple hulls.

The bean family has lots of good crops for rotations. They are easy to save seed from to plant again next year. Lima beans, bush and pole beans, pintos, black turtles and numerous other varieties will have a place in the garden.

An interesting member of this family is the peanut. A long shoot comes out of the flower and buries into the soil, where it makes the peanut. So, although it forms underground, the peanut if actually a seed made by the pollination of flowers.

Some trees and bushes are legumes, such as locusts and redbuds. Legumes have distinctive flowers, pods with their seeds, and compound leaves. The soil around legumes is richer than the surrounding soil.

Crop Choices. — Besides rotating crops and animals from field to field, farmers grow a wide variety of crops every year. Diversity is a sign of health. A farmer's success depends upon the choice of crops.

The farm's location, microclimate, soil condition, water holding capacity, air flow, sunshine aspect, wildlife pressure, and market access all play critical roles in the determination of what to grow. New farmers pay close attention to what crops their neighbors are growing, and the timing and methods they use. Different kinds of equipment are used for growing various crops.

Some crops, such as corn and wheat, are widely grown, and others, like citrus or peaches, are only raised in specific regions. The plains states grow grains, potatoes, and cabbage. Southern states, with their longer seasons, grow cotton, tobacco, and sweet potatoes. In a Mediterranean climate they raise wine grapes. Most crops can be grown in other places but only thrive when they are in certain climates.

Alfalfa requires a deep, rich soil. Clover is used in a more shallow soil. In the low fertility south they grow sirica lespedeza, but as the soil improves they can grow clover.

Garden and fruit crops need good air circulation to dry them off in the morning and supply fresh air. Many crops like a good view, such as a hill top with lots of sun and good air drainage. But in the windy Midwest a wind break of conifer trees may be needed to protect the plants from excess airflow.

Plant growth is a function of sunlight. The longer a plant gets sun, the more photosynthesis can happen and the more it can grow. Tall hills or trees in the east or west blocking the morning or evening sun will compromise sun-loving plants. A southern slope will be warmer and dry out quicker than a northern slope.

Deer and other wildlife limit what can be grown without protection from them. Fencing may be required to grow crops they can destroy. Because deer do not have depth perception, a short three-dimensional fence will suffice. This is usually a couple of wires surrounding the field with another wire that's electrified encircling the first fence about four feet out. Otherwise, an eight foot tall fence will be necessary.

The location of markets determines crop choices, especially perishable ones. All of these factors play into the farmer's decision on what to raise. While experimenting with new or different crops should be encouraged, there is usually a reason a certain region grows certain crops and animals.

Planting— Primary tillage requires more horsepower than subsequent cultivation. Once the land is plowed and settled it should be free of growing vegetation. A farmer needs a stale seed bed to get a jump ahead of the weeds. Then furrows are made at a distance apart that will make cultivation easy.

Seeds must have firm contact with the soil, so they are pressed into the ground by stepping on them or rolling them with a culitpacker. Loose, fluffy soil is ideal for growing plants as weeds won't sprout, but the crop seed needs good soil contact for absorbing moisture to sprout.

Dry soil is then thrown over the seed in the furrow. The farmer never works the soil when it is wet. Moisture from below will rise by capillary action to sprout the seeds, and the dry soil on top checks evaporation.

Long season tropical plants, like tomatoes, peppers, eggplants, and sweet potatoes, are often started early in a cold frame. These are simple boxes sloping slightly to the south with removable window sashes. For extra warmth they are sometimes built on a horse manure heap, or are dug out an extra foot deep and filled with horse manure. The manure from horses is not as digested as cow manure and hence heats up more readily, thus turning the cold frame into a hotbed.

Early crops of lettuce and cabbage family plants are also often started in cold frames. All of these crops transplant easily. For most other crops, the seeds are just sown directly into the field.

Plants are pulled bare root from the cold frames and their roots are dampened immediately. They are held at the stem so oils from the hand don't touch the roots. Sunshine will wilt them, so shade is needed once they are pulled. The addition of sand and compost to the soil makes pulling up the plants easier.

Cold frame on top of warm horse manure

A hole is made in the furrow and a pint or two of water is poured into it. The plant is set into the mud and dry earth is pulled over it. Plants can be set and then watered, but dry earth is still pulled around the plant so that the wet earth does not have a chance to dry out. Soil moisture goes into the air if the soil is moist, muddy, or caked, and the tiny plant stem can be choked as this soil dries and hardens.

Cover crops and grains are sown by broadcasting the seeds. A steady, sweeping arch is made with a handful of seeds every few steps. A harrow or cultipacker is used to cover them up.

Berry and orchard land should be subsoiled and cover cropped a year before planting to get the soil loose and well-drained. Holes are dug and the broken roots are pruned off. Dry soil is firmed around the roots, they are watered, and then more dry soil is pulled on top. They should not be set any deeper than they were in the nursery.

Market gardeners must be careful not to plant when they have crops needing weeding. They don't want a week of wet weather to delay weeding, and it is better to not have new seeds in waterlogged soil. It is preferable to tend well what is already planted before sowing more.

Planting the same crop three or four weeks later will ensure a succession of harvests later. This is often done with corn, cucumbers, beans, and summer squash. Otherwise summer cover crops, such as buckwheat or cowpeas, are sown to smother grass and weeds and improve the soil after the crop is finished.

Animals— All farms have animals. As we have seen, the ruminants play an extremely important role, by being able to make the land more fertile while getting their own feed from it. Other animals, like horses, pigs and poultry, cannot do this. They have other roles to play on the farm.

Animals do not get nutrition directly from the atmosphere and soil like plants do. They eat plants, breaking down the nutrients indigestion and then building them back up in their bodies. They need carbohydrates in the forms of starch, sugars and fats, and also nitrogen-rich foods in the form of protein.

Small farms or gardens also have animals. Besides the wildlife routinely visiting, we have to count the insects, earthworms, and the soil microorganisms as animals. The ability of all animals to transform plant material is indispensable for decomposing wastes and growing more plants. It all works together symbiotically. Gardens generally require bringing in animal wastes from farms to make compost.

Cows are the ultimate farm animal. They have been with humans at least since recorded history and were bred to be primarily for meat, milk, or work animals. The beef breeds include Angus, Hereford, Charlois, and Shorthorns. Dairy breeds include Jersey and Guernsey, which produce milk with a higher butterfat content, and Holstein, Ayrshire and Brown Swiss, which produce a larger quantity of milk. When crops are grown and fed to cattle, and the wastes returned to the cropland, beef and milk can be exported without depleting the soil's fertility.

Cows
Showing the Dairy Form
Showing the Beef Form

Sheep breeds are divided into those grown for wool, like Merino and Southdown, and hairless sheep grown for meat. Because of problems with foot rot and parasites they are raised in hill top areas that do not have excess moisture.

Goats also have meat and hair breeds. Brush goats are often used to clean up areas that have grown up in primary forest species. They are capable of eating a much wider variety of plants than cattle. Both sheep and goats also have breeds for milking.

Horses and mules were commonly raised when they supplied transportation and power. Their hooves pack the soil, so farmers rarely keep them except for recreation. Because they are not ruminants, their manure is better if it is decomposed with straw, wood chips, or other carbon rich material before being spread on the fields.

Pigs and poultry are the animals that live off the farm's waste products. They supply meat and eggs at little expense to the farmer, who lets them have what nobody else wants. Chickens and other poultry follow the larger animals around, finding food in their wastes and scratching what's left into the soil.

Pigs are often confined and feed is brought to them. This prevents their destructive nature of rooting up soil on the rest of the farm. Pigs can kill trees and are used to clear forest land for pasture.

The more often ruminants are moved the better it is for them and the soil. Imagine 30 head of cattle on 100 acres all the time, called set grazing. They would quickly eat their favorite plants and the other plants would then dominate the pasture. As soon as their favorite plants grew back a little, they would eat these plants. This would gradually eliminate the best food plants from the pasture.

On the other hand, if the pasture was divided into 100, one acre paddocks, the 30 head of cattle will pick the one acre down evenly. The next day they are allowed to go into a new one acre paddock. The first acre will then have 99 days to recover before the cattle go in there again, allowing for regrowth of their favorite plants and a deepening of the topsoil because of no animal impact on it.

More paddocks and moving more times each day would be even better. A farmer has to determine how often to move the animals. It will be different during the various seasons, with quicker rotations while the plants are growing faster. Animals intensively grazed in frequent rotations are healthier, their coats shine, and they gain more weight, but it is a trade-off the farmer makes for the time and energy spent.

A good way to remineralize the farmland is to give minerals to the animals, who will distribute them over the fields in their droppings. They know to lick the mineral salts that are missing from the farm's soil. Remember that everything in nature is interconnected. Even in a small garden, the earthworms and other soil life are moving nutrients around, with the help of insects, reptiles, birds and mammals. Soil is never without animals.

Afterword

Rudolf Steiner was not a farmer, but he understood the common farming wisdom presented in the appendix. Foretelling the dangers of using artificial fertilizers, especially nitrates, he called it an absolute general law that products from fields treated with them would lose their nutritional value. His recommendations to rely on manures, compost, and other traditional methods were a lone cry in the wilderness in 1924.

Some of his ideas are hard to swallow, while other make common sense. The guiding lines and innovative ways to think about farming come from a very broad perspective and he urged us to experiment with them. For example, most farmers get more successful results with the preparations in lectures four and five than they do with the ones in lecture six. His advice is to observe, contemplate, recognize conditions, develop our own insights, and to look at our farms as individualized, self-sufficient living organisms.

He also suggested avoiding chemistry on our farms. So began the movement for organic food and farming, and it continues gaining momentum today. Biodynamic products routinely rate the highest of quality worldwide. As the following quote indicates, people have long understood the necessity of crop rotations, manure, ashes, not plowing land and resting the fields in cover crops. Steiner looked at this peasant wisdom from the scientific viewpoint of biochemistry, and developed the Agriculture Course with human nutrition as the ultimate goal of farming. One hundred years have only amplified the importance of his insights.

Still, by the rotation of crops
You lighten your labor,
Only hesitate not to enrich
The dried up soil with dung
And scatter filthy ashes
On fields that are exhausted.
So, too, are the fields rested
By a rotation of crops
And unplowed land in the meanwhile
Promises to repay you.

-The Georgics of Virgil
30 B.C.

www.ingramcontent.com/pod-product-compliance
Lightning Source LLC
Chambersburg PA
CBHW072010290426
44109CB00018B/2201